Advances in Urban Stormwater and Agricultural Runoff Source Controls

NATO Science Series

A Series presenting the results of scientific meetings supported under the NATO Science Programme.

The Series is published by IOS Press, Amsterdam, and Kluwer Academic Publishers in conjunction with the NATO Scientific Affairs Division

Sub-Series

I. **Life and Behavioural Sciences**	IOS Press
II. **Mathematics, Physics and Chemistry**	Kluwer Academic Publishers
III. **Computer and Systems Science**	IOS Press
IV. **Earth and Environmental Sciences**	Kluwer Academic Publishers
V. **Science and Technology Policy**	IOS Press

The NATO Science Series continues the series of books published formerly as the NATO ASI Series.

The NATO Science Programme offers support for collaboration in civil science between scientists of countries of the Euro-Atlantic Partnership Council. The types of scientific meeting generally supported are "Advanced Study Institutes" and "Advanced Research Workshops", although other types of meeting are supported from time to time. The NATO Science Series collects together the results of these meetings. The meetings are co-organized bij scientists from NATO countries and scientists from NATO's Partner countries – countries of the CIS and Central and Eastern Europe.

Advanced Study Institutes are high-level tutorial courses offering in-depth study of latest advances in a field.
Advanced Research Workshops are expert meetings aimed at critical assessment of a field, and identification of directions for future action.

As a consequence of the restructuring of the NATO Science Programme in 1999, the NATO Science Series has been re-organised and there are currently five sub-series as noted above. Please consult the following web sites for information on previous volumes published in the Series, as well as details of earlier sub-series.

http://www.nato.int/science
http://www.wkap.nl
http://www.iospress.nl
http://www.wtv-books.de/nato-pco.htm

Series IV: Earth and Environmental Series – Vol. 6

Advances in Urban Stormwater and Agricultural Runoff Source Controls

edited by

Jiri Marsalek
National Water Research Institute,
Burlington, Ontario, Canada

Ed Watt
Queen's University,
Kingston, Ontario, Canada

Evzen Zeman
DHI Hydroinform a.s.,
Prague, Czech Republic

and

Heiko Sieker
Ingenieurgesellschaft Prof. Dr. Sieker mbH,
Berlin, Germany

Kluwer Academic Publishers

Dordrecht / Boston / London

Published in cooperation with NATO Scientific Affairs Division

Proceedings of the NATO Advanced Research Workshop on
Source Control Measures for Stormwater Runoff
St. Marienthal-Ostritz, Germany
8–12 November 2000

A C.I.P. Catalogue record for this book is available from the Library of Congress.

ISBN 1-4020-0153-3 (HB)
ISBN 1-4020-0154-1 (PB)

Published by Kluwer Academic Publishers,
P.O. Box 17, 3300 AA Dordrecht, The Netherlands.

Sold and distributed in North, Central and South America
by Kluwer Academic Publishers,
101 Philip Drive, Norwell, MA 02061, U.S.A.

In all other countries, sold and distributed
by Kluwer Academic Publishers,
P.O. Box 322, 3300 AH Dordrecht, The Netherlands.

Printed on acid-free paper

TABLE OF CONTENTS

CHAPTER 5   SOURCE CONTROLS IN RURAL CATCHMENTS

CHAPTER 6   SOURCE CONTROL IMPLEMENTATION

PREFACE

Large rainfall and snowmelt events cause numerous water management challenges in both urban and rural areas. In urban areas, these events cause stormwater runoff, which results in a variety of impacts on receiving waters, including flooding, geomorphologic changes, pollution, ecosystem degradation and impairment of beneficial water uses. To prevent or mitigate such impacts, stormwater management has been introduced and implemented through a system of management measures, which are also referred to as 'best management practices' (BMPs). Among BMPs, source controls are particularly widely accepted, because they are designed to prevent or mitigate the stormwater impacts by applying controls at or near the source. Consequently, they are generally highly cost-effective, and in some cases represent the only practical solution (e.g., eliminating toxic substances releases, rather than containing and removing such substances after their dispersal in the environment).

Concerns about stormwater runoff in agricultural (rural) areas are similar to those in urban areas, and include the risk of flooding, sediment erosion and delivery to receiving waters, and washoff and transport of agrochemicals and pathogens. In recent years, a great deal of attention has been paid to mitigation of these problems by source controls. It has been suggested that by proper land management, catchment runoff could be reduced (or even eliminated for small events), and reduced runoff flows would then contribute to lower flood risks, soil erosion, and export of agrochemicals and pathogens.

In view of this great interest in source controls for stormwater runoff, and particularly in using these measures to mitigate flood risks, we proposed a research workshop on this subject and applied for NATO sponsorship. Thus, the main objective of the NATO Advanced Research Workshop on 'Source Control Measures for Stormwater Runoff' was to review the state of the art in source controls of stormwater runoff in both urban and agricultural areas, with respect to the management of wet-weather flows. For this workshop, we have used a broad definition of source controls as the measures which are designed to control the generation of, and entry of pollutants into, stormwater runoff, with emphasis on non-structural and semi-structural measures applied at or near source.

After receiving the NATO support grant, we recruited keynote speakers and workshop participants, finalised the workshop programme, and held the workshop at the International Conference Centre in St. Marienthal-Ostritz, Germany, with more than 50 invited experts from 19 countries in attendance. Extensive experience of workshop participants in this field is reflected in the workshop proceedings, which include 30 selected papers. Whenever trade, product or firm names are used in the proceedings, it is for descriptive purposes only and does not imply endorsement by the Editors, Authors or NATO.

x

Only the formal workshop presentations are reflected in the proceedings that follow. Besides these papers and extensive discussions, there were many other ways of sharing and exchanging information among the participants, in the form of new or renewed collaborative links, professional networks and personal friendships. The success of the workshop was acknowledged by workshop participants in the evaluation questionnaire. For this success, we would like thank all who helped stage the workshop and produce its proceedings, and particularly those who are listed in the Acknowledgement.

Jiri Marsalek
Burlington, Ontario, Canada

Ed Watt
Kingston, Ontario, Canada

Evzen Zeman
Prague, Czech Republic

Heiko Sieker
Berlin, Germany

ACKNOWLEDGEMENT

This Advanced Research Workshop (ARW) was directed by Dr. Jiri Marsalek, National Water Research Institute (NWRI), Environment Canada, Burlington, Canada, and Dr. Evzen Zeman, DHI Hydroinform a.s., Prague, Czech Republic. They were assisted by two other members of the workshop Organising Committee, Prof. W.E. Watt, Queen's University, Kingston, Ontario, Canada, and Dr. H. Sieker, Ingenieurgesellschaft Prof. Dr. Sieker mbH, Berlin, Germany. The ARW was sponsored by NATO, in the form of a grant, and by employers of the members of the Organising Committee, who provided additional resources required to prepare the workshop and its proceedings. Special thanks are due to Dr. A.H. Jubier, Programme Director, Environmental and Earth Science & Technology, NATO, who provided liaison between the workshop organisers and NATO, and personally assisted with many tasks.

The early preparatory work was conducted by the DHI Hydroinform team (Evzen Zeman and Pavlina Nesvadbova) and the National Water Research Institute team (Jiri Marsalek and Quintin Rochfort). All local arrangements were carried out by the local organising committee, headed by Dr. Heiko Sieker, with assistance from Brigitte Leipold and Harald Sommer. Finally, the proceedings typescript was compiled by Karen MacIntyre (Queen's University, Kingston, Canada) and Quintin Rochfort (National Water Research Institute, Burlington, Canada). Special thanks are due to all these contributors and, above all, to all the participants, who made this workshop a memorable interactive learning experience for all involved.

LIST OF PARTICIPANTS

Directors

Marsalek, J   National Water Research Institute
867 Lakeshore Road, Burlington, Ontario, L7R 4A6
CANADA

Zeman, E.   DHI Hydroinform a.s.
Na Vrsich 5, 100 00 Praha 10
CZECH REPUBLIC

Key Speakers

Bournaski, E.   Bulgarian Academy of Sciences, IWP
Acad. G. Bontchev Street, Block 1, BG -1113 Sofia
BULGARIA

Watt, W.E.   Dept. of Civil Engineering
Queen's University, Kingston, ON K7L 3N6
CANADA

Kuby, R.   DHI Hydroinform a.s.
Na Vrsich 5, 100 00 Praha 10
CZECH REPUBLIC

Metelka, T.   Hydroprojekt a.s.
Taborska 31, 140 00 Praha 4
CZECH REPUBLIC

Pryl, K.   DHI Hydroinform a.s.
Na Vrsich 5, 100 00 Praha 10
CZECH REPUBLIC

Mikkelsen, P.S.   Dept. of Env. Science and Engineering
Tech. University of Denmark, Bldg 115, DK-2800 Kongens
Lyngby
DENMARK

Raimbault, G.   LCPC
BP 4129, 44341 Bouguenais Cedex
FRANCE

Sieker, F.   University of Hannover
Heinrich Beensen-Str. 1, D-30926 Seelze
GERMANY

Sieker, H.	Ingenieurgesellschaft Prof.Dr. Sieker mbH Berliner Strasse 71, D-15366 Dahlwitz-Hoppegarten GERMANY
Petrovic, P.	Water Research Institute Nab. gen. Svobodu 5, 812 49 Bratislava SLOVAKIA
Geldof, G.	University of Twente, Faculty of Tech. and Management P.O. 217, 7500 AE Enschede THE NETHERLANDS
Crabtree, R.W.	WRc plc Frankland Road Blagrove, Swindon SN5 8YF UK
Earles, T.A.	Wright Water Engineers, Inc. 2490 West 26th Ave., Suite 100A, Denver, CO 80211-4208 USA
Quigley, M.	GeoSyntec Consultants 532 Great Road, Acton, MA 01720 USA

Other Participants

Vaes, G.	Hydraulics Laboratory, University of Leuven de Croylaan 2, B-3001 Heverlee BELGIUM
Tenev, B.	Res. Inst. for Land Reclamation and Hydraulic Engineering 136 Tzar Boris III Blvd., BG-1618 Sofia BULGARIA
Baca, V.	Povodi Vltavy a.s. Holeckova 8, 150 21 Praha 5 CZECH REPUBLIC
Baloun, J.	Povodi Vltavy a.s. Litvinovicka 5, 371 21 Ceske Budejovice CZECH REPUBLIC
Biza, P.	Povodi Moravy a.s. Drevarska 11, 601 75 Brno CZECH REPUBLIC

rasek, V.   Povodi Labe a.s.
                             Vita Nejedleho 951, 500 03 Hradec Kralove 3
                             CZECH REPUBLIC

aichter, T.   Hydroprojekt a.s.
                             Taborska 31, 140 00 Praha 4
                             CZECH REPUBLIC

rax, P.   Technical University of Brno, Institute of Municipal Engineering
                             Zizkova 17, 602 00 Brno
                             CZECH REPUBLIC

uma, A.   Povodi Moravy a.s.
                             Drevarska 11, 601 75 Brno
                             CZECH REPUBLIC

Bronstert, A.   Inst. of Geo-Ecology, Potsdam University
                             PO-Box 60 15 53, 14415 Potsdam
                             GERMANY

Ostrowski, M.   Darmstadt University of Technology
                             Petersenstr. 13, D 64287 Darmstadt
                             GERMANY

Remmler, F.   Institute for Water Research
                             Zum Kellerbach 46, 58239 Schwerte
                             GERMANY

Schmidt, W.   Saechs. Landesanstalt fuer Landwirtschaft
                             Gustav-Kuehnstr. 8, D-04159 Leipzig
                             GERMANY

van der Ploeg, R.   Institute of Soil Science, University of Hannover
                             30419 Hannover
                             GERMANY

Wilcke, D.   University of Hannover, Institute of Water Resources
                             Appelstr. 9a, D-30167 Hannover
                             GERMANY

Aftias, E.   National Technical University of Athens
                             5, Heroon Polytechniou str., 15780 Athens
                             GREECE

Balint, G.	Hydrological Institute, VITUKI Kvassay u. 1, Budapest H-1095 HUNGARY
Giulianelli, M.	IRSA via Reno 1, 00198-ROMA ITALY
Andersen, T.	Statkraft Groner AS PO Box 400, N1327 Lysaker NORWAY
Dziopak, J.	Water Supply and Sewage Systems, Rzeszow University of Technology ul. Powstancow Warszawy 6, 35-959 Rzeszow POLAND
Sowinski, M.	Inst. of Environmental Engineering, Poznan University of Technology Piotrovo St. 5, 60-965 Poznan POLAND
Woloszyn, E.	Faculty of Environmental Engineering, Technical University of Gdansk ul. Narutowicza 11/12, 80-952 Gdansk POLAND
do Ceu Almeida, M.	LNEC/DH/NES Av. Do Brasil 101, 1700-066 Lisbon PORTUGAL
Matos, R.	LNEC/DH/NES Av. Do Brasil 101, 1700-066 Lisbon PORTUGAL
Sokac, M.	Faculty of Civil Engineering, Slovak Technical University Radlinskeho 11, 813 68 Bratislava SLOVAKIA
Sztruhar, D.	Faculty of Civil Engineering, Slovak Technical University Radlinskeho 11, 813 68 Bratislava SLOVAKIA

rnic, J. Okolje Consulting DOO
Prvomajska 41, 63000 Celje
SLOVENIA

lalmqvist, P.-A. Chalmers University of Technology, Urban Water
S-412 96 Gothenburg
SWEDEN

vensson, G. Chalmers University of Technology, Water Environment
Transport
S-412 96 Gothenburg
SWEDEN

teiner, M. EAWAG (SWW)
Ueberlandstr. 133, 8600 Duebendorf
SWITZERLAND

utler, D. Imperial College of Science, Technology and Medicine
Department of Civil Engineering, E&WR Section
London SW7 2BU
UK

oychenko, S. Dept. of Meteorology and Climatology
Kyiv Taras Shevchenko University
st. Vassilkivska 90, Kyiv 022
252022 UKRAINE

Voloshchuck, V. Dept. of Meteorology and Climatology
Kyiv Taras Shevchenko University
st. Vassilkivska 90, Kyiv 022
252022 UKRAINE

Iovotny, V. Institute for Urban Environmental Risk management
Marquette University
PO Box 1881, Milwaukee, WI 53201-1881
USA

REVIEW OF STORMWATER SOURCE CONTROLS IN URBAN DRAINAGE

J. MARSALEK
National Water Research Institute, 867 Lakeshore Rd, Burlington, ON L7R 4A6, Canada

1. Introduction

Discharges of urban stormwater into receiving waters may cause a variety of impacts leading to flooding, geomorphological changes, pollution, ecosystem degradation and impairment of beneficial water uses [1, 2]. To prevent or mitigate such impacts, stormwater management has been introduced and generally implemented through a system of management measures, which are also referred to as 'best management practices' (BMPs). Among BMPs, source controls are particularly widely accepted, because they are designed to prevent or mitigate the impacts by controls at or near the source. Consequently, they are generally highly cost-effective, and in some cases represent the only practical solution (e.g., eliminating toxicant releases, rather than containing and removing such substances after their dispersal in the environment). Although the need for source controls is widely recognised, in many cases, their performance cannot be verified and may suffer from local limitations. Consequently, stormwater management treatment trains are designed to meet project objectives by combining both source controls as well as more robust structural BMPs [3].

The terminology on stormwater source controls is not standardised and this term is used differently by various drainage professionals. A recent U.S. engineering manual of practice [2] defines source controls as "practices that prevent pollution by reducing potential pollutants at their source before they come into contact with stormwater...". Other sources present a broader definition of source controls as measures "limiting changes to the quantity and quality of stormwater at or near source" [4]. Recognising that this paper should serve a broad audience, urban stormwater source controls were defined here as the measures designed to control the generation of, and entry of pollutants into, stormwater runoff, with emphasis on non-structural and semi-structural measures applied at or near source.

The primary purpose of this review was to establish a framework for discussions of urban stormwater source controls at the NATO Advanced Research Workshop on this subject, which was held at St. Marienthal, Germany in November, 2000. The review starts with source control measures affecting both stormwater quantity and quality, which are followed by the measures enhancing stormwater quality.

J. Marsalek et al. (eds.), Advances in Urban Stormwater and Agricultural Runoff Source Controls, 1–15.
© *2001 Kluwer Academic Publishers. Printed in the Netherlands.*

2. Source Control Measures Reducing Stormwater Quantity

Source control measures reducing stormwater quantity are generally those which reduce the extent of impervious areas; divert runoff from impervious areas onto pervious areas; enhance hydrologic abstractions on natural or man-made surfaces by detention/retention, infiltration and evaporation; and, reduce stormwater flows by storage and reuse. While these measures primarily control runoff quantity, they also improve stormwater quality by immobilising stormwater pollutants on catchment surfaces, or diverting them to soils, or groundwater. A brief overview of selected measures follows.

2.1. LAND USE PLANNING AND MANAGEMENT PRACTICES

The purpose of these practices is to minimise the potential for runoff and pollutant export generation as a result of land development. They represent a conservation approach to land development, employing land use capability assessment mapping and land use zoning and management practices consistent with minimising the effects of urbanisation on receiving waters and integrating natural drainage into new urban developments [5]. As much as possible, runoff flows through natural channels or vegetated swales, enhancing runoff filtration and infiltration, into a network of wetlands and ponds, which reduce peak flows. Furthermore, buffers for streams and wetlands are provided. This approach, which is widely accepted in new developments, has numerous benefits (cost savings, improved market value and resource utilisation, environmental benefits), but also environmental and social constraints/ limitations, which are encountered in its implementation [5]. For implementation and enforcement, land use ordinances are needed. As is the case with most source controls, public support is needed to gain acceptance for land use policies. Some land use restrictions may not be politically feasible [2].

2.2. LOT-LEVEL SOURCE CONTROLS

Lot-level source controls are small scale measures serving to reduce runoff flows and sometimes enhance their quality, control erosion, and enhance groundwater recharge [6]. Such results can be achieved by lot grading allowing water ponding and enhancing infiltration; rooftop detention/retention storage [6]; disconnection of roof leaders; parking lot storage; runoff conveyance by grassed swales and through filter strips; stream and corridor buffer strips [7, 8], storage in underground structures; runoff infiltration in soakaway pits, trenches and perforated pipes [9]; and, runoff storage and reuse [10, 11].

Lot-level source controls are very diverse and can be effective, if applied throughout the catchment, as often done in new developments. There are examples, e.g. collecting rainwater in rain barrels for watering (in the Toronto area), where these measures served to promote public education on, and participation in, stormwater management. On the other hand, some of these measures are susceptible to tampering, which may reduce their effectiveness without inspection programs. Details of these measures can be found in BMP manuals [6, 7]; selected ones are described below.

Lot grading is used to slow down runoff (e.g., by reducing, slopes from 2% to 0.5%, where permitted by the local municipality) and create ponding areas, generally less than 0.1 m deep. Roof leaders discharge onto property, or into soakaway pits. Flat roof storage can be created by drainage hoppers fitted with weirs; parking lot storage is created by sewer inlet control devices [6].

Stormwater biofiltration by grass filters and swales is used commonly to reduce runoff volume by infiltration and enhance runoff quality by such processes as settling, filtration, adsorption and bio-uptake. Vegetated filter strips are feasible in low density developments with small contributing areas and diffuse runoff, suitable soils (good sorption), and lower groundwater tables. Good design parameters include a flow depth of < 50-100 mm, length > 20 m, and slopes 1-5% [7].

Swales are shallow grassed channels functioning in a similar way as vegetated biofilters. Good design features include the bottom width > 0.75 m, mild slopes (< 1%), small contributing areas (< 2 ha), and an adequate length L > 60 m. Swales are best suited for small areas with permeable soils and low groundwater tables [7]. In a summary of U.S. experience with swales, Schueler *et al* . [8] noted that swales offered mixed performance in removal of suspended solids and attached pollutants, and low removal of solubles. In-swale biofilters improved their performance. Swales have good longevity, if well maintained. Compared to conventional curb & gutters, they are less expensive to build, but require more land. Leaching from culverts and fertilised lawns may increase the presence of trace metals and nutrients [8].

In a recent design, the stormwater swale is underlain by a gravel infiltration trench (see Fig. 1) with a throttled drain pipe (the so-called MR system). In this arrangement, stormwater infiltrating through an active soil layer is 'pretreated' before entering a gravel trench, and drains via a drain pipe discharging into a manhole with a flow throttle [12].

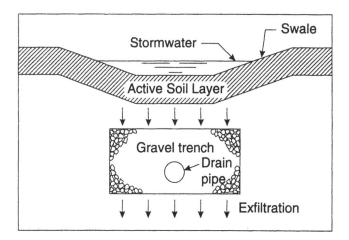

Figure 1. Cross-section of a swale with underdrain (after [12]).

Infiltration facilities serve to reduce the volume and rate of runoff, reduce pollutant transport and recharge groundwater. They are designed in various forms, including wells (pits), trenches, basins, and perforated pipes and drainage structures (catchbasins, inlets, and manholes), often equipped with some pre-treatment measures, an under-drain and bypass [6, 7, 13]. Porous pavement represents another infiltration measure for reducing surface runoff and treating infiltrated stormwater [14, 15]. All these structures reduce the volume of runoff by allowing it to infiltrate into the ground. Since the infiltrating runoff contains some pollutants, infiltration facilities also control pollution export from drained areas. The use of infiltration is generally feasible in small residential areas with low risk of groundwater contamination, soils with good percolation rates, and deeper groundwater or bedrock. Septic tanks and building foundations have to be avoided. In general, infiltration structures can be very cost effective [16]; the main difficulties with their applications include potential contamination of groundwater and uncertain longevity of these structures [7, 8].

Water quality inlets (also referred as oil/grit separators) provide partial stormwater treatment by sedimentation and skimming of floatables (oil), and thereby pretreat stormwater before its discharge into sewers. They were recommended for applications in areas with high pollution/spill potential, including parking lots, and commercial or industrial developments [6], where they were designed to provide a permanent storage of 15 m^3/impervious ha. A number of designs indicate good potential for removing coarse solids (sand) and containing free oil spills [6]. Recent findings indicate that in cold climates, their treatment effectiveness may be reduced by accumulation of chlorides [17]. For areas with sensitive receiving waters, a more complex device, referred to as the multi-chambered treatment train (MCTT), was recently developed by the US EPA [18]. It comprises a catchbasin/grit chamber, followed by a two-chambered tank, with flash aeration, enhanced settling (tube of plate settlers), and a slow, mixed media filter (sand and peat). Other inlet designs employ oil sorbents to reduce oil entry into storm sewers.

Another treatment process applicable at a small scale (e.g., an industrial property) is filtration. Stormwater sand filters were introduced in the USA and found effective in removing pollutants [19]. They need to be regularly regenerated by backwashing [7], or simply by raking the surface layer; they may get clogged by formation of a biofilm [19]. Good designs may serve up to 5 ha, use a sand layer of 0.5 m, operate with a hydraulic head of 0.6-1.0 m (higher heads compact sand), and are equipped with a collector of the filtrate and an overflow/bypass structure [19]. Biofilters (i.e., coarse medium filters with biofilm grown on granular surfaces) were also tested and show good promise for removal of dissolved heavy metals from stormwater [20,21]. Porous pavements with subsurface storage are used for stormwater filtration and quality enhancement by sorption on the fill material. Pratt *et al* . [15] and Raimbault *et al* . [22] demonstrated good treatment efficiency of these structures.

3. Source Controls Enhancing Stormwater Quality

Source controls enhancing stormwater quality are generally policies and related measures which reduce or eliminate entry of numerous pollutants [23] into stormwater. These measures are generally designed to promote the prevention of stormwater pollution by various activities conducted by the public, municipalities and small businesses. Many such measures are described in great detail in the literature [2, 4, 5, 24] and are recommended for application in both existing and new urban developments. Selected controls are discussed below.

3.1. PUBLIC EDUCATION, AWARENESS AND PARTICIPATION

Public awareness, education and empowerment programs (PAEEP) are essential for planning, design and acceptance of new stormwater facilities, by promoting identification and understanding of drainage problems and their remediation, identifying responsible parties and past efforts, promoting community ownership of these problems and solutions, promoting changes toward environmentally responsible practices and behaviour of both individuals and corporations, and, integrating public feedback to program implementation [2, 4]. Consequently, PAEEP is recognised as an appropriate strategic tool for reducing stormwater pollution. The success of most source control measures depends on public education and participation, particularly if links between individual behaviour and water quality are established, individuals are targeted both at home and work, and the whole community is mobilised. At the same time, it is difficult to assess the effectiveness of these programs.

To develop an effective PAEEP, environmental issues, audience, behaviour targeted, and best ways to achieve an improved environment must be understood. Such a program can be developed through the steps listed below [4].

- Define and analyse the problem (the sources of pollution, who impacts on them).
- Identify stakeholders (commercial business, industry, land holders and residents, school/youth groups, and municipal staff).

- Know your target group - establish a complete profile, develop the best methods of communication.
- Set objectives - informative messages, emotional messages, responsibility messages, empowering messages, action messages (clear simple, language; technical sound statements; break up concept into simple statements).
- Design your methods - determine a mixture techniques for the target group.
- Form action plans and timelines - identify costs, funding sources and trim the project to fit resources.
- Monitor and evaluate - collect information and records to see how effective this is, recognising that there may be a lag in public response.

Elements of the PAEEP program include printed materials, media, signs, community programs, displays, community water quality monitoring programs, launches, local action committees and groups, advisory groups, consumer programs, business programs, and school education programs. Storm drainage system signs, with prohibitive language and icons to discourage polluting, were found particularly effective [2]. To reduce costs, this activity should be implemented mostly with a volunteer workforce, but in high traffic zones, municipal staff should be employed.

Awareness and education are implemented through public meetings, open houses, tours of facilities, and visual displays at stormwater management sites. In this process, concerned citizens/environmentalists groups are formed which then actively engage in environmental projects, including organised cleanup and publicity campaigns, and help involve schools [2, 4].

3.2. MODIFIED USE, RELEASES AND DISPOSAL OF CHEMICALS ENTERING STORMWATER

These measures employ planning, and environmental and building ordinances and regulations to reduce releases of harmful chemicals into stormwater, generally by changed behaviour with respect to engaging in some activities, use of certain products, and handling and disposal practices [5]. These measures may be introduced at various administrative levels, e.g., the phasing of lead out of gasoline was done at the national level, and similar benefits follow from efforts of clean air agencies [25, 26]. At the municipal level, environmentally safe use and management of household materials and chemicals, including pesticides, solvents, oils, fertilisers, and antifreeze are promoted to minimise the potential for washoff or leaching or indirect disposal into the stormwater system [5, 25, 26]. Specific measures include reduction in activities causing chemical releases (e.g., driving), safe use of chemicals (e.g., proper applications of garden chemicals without excessive releases), substituting safe alternative products for toxicants, and proper handling and disposal of chemicals [2].

Use of vehicles can be reduced through public education and establishment of trip reduction programs in government offices and larger business. These activities should be pursued in alliance with clean air agencies, may require extra staff to coordinate, and their success may be limited by the lack of cooperation, concerns about the loss of convenience and relative costs [2].

7

Many garden chemicals are potentially toxic and require environmentally safe handling. Examples include proper use of fertilisers and pesticides, without over-application and spread on directly connected impervious areas [26]. At the municipal level, there is a growing movement to reduce of eliminate the use herbicides on public lands and minimise the use of salt in winter road maintenance [27].

Road salts are used extensively in cold climates for winter road maintenance, as shown for selected Canadian cities in Fig. 2. Such use has been of environmental concern for some time and this concern has led to placing road salts on the (Canadian) Priority Substances List [28]. While there are environmentally safer alternative deicers (e.g., calcium magnesium acetate, CMA), they are very costly. Environmental impacts can be partly mitigated by reduced salt usage resulting from such measures as pre-wetting salt, using salt brine, using calibrated spreaders, improved operational practices (sensible salting), keeping salt under cover, and pavement heating or modification by hydrophobic or icephobic coatings [27]. More widespread use of alternative products may require educating the public and municipal staff about such alternatives, and advising purchasing departments [2].

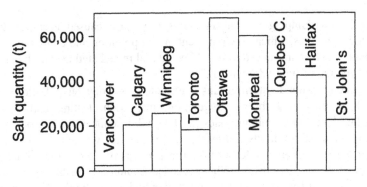

Figure 2. Road salt use in selected Canadian cities during winter 1997-98 (after [28]).

Household hazardous waste, that could end up in stormwater, needs to be collected, usually in conjunction with solid waste collection, and properly disposed of. A proper collection plan has to be developed and address such issues as optimal methods of collection and its frequency. These programs are subject to existing regulations, may be relatively costly, require trained operators, appropriate laboratory and detection equipment, and good record-keeping. Successful programs rely on public education and participation, informing the public about hazardous materials, safe alternatives, proper storage and disposal, and promoting reuse and recycling. Relatively high costs limit more widespread use of these programs [2]. Used oil recycling is also a form of hazardous waste management. To avoid improper disposal, used oil collection and recycling should be set up and operated by private contractors. These programs are subject to many regulations and may be limited by the lack of reliable licensed used oil recyclers [2].

8

3.3. ENFORCEMENT OF SEWER ORDINANCES

Many pollution problems arise from violations of sewer ordinances and the lack of their enforcement. Detection of problems and implementation of corrective measures are key activities in mitigating these problems. The types of activities addressed here include illegal dumping control, removal of contaminated sediment from sewers, prevention of illicit connections, illicit connection detection and removal, and control of leaking sanitary sewers.

Illegal dumping should be fought by a two-pronged approach - (a) educating the public about the harm caused by these activities, and (b) instituting ordinances to detect and correct dumping, and enforce penalties. Municipal codes should prohibit illegal dumping and additional staff may be needed to detect and investigate illegal dumping, and coordinate public education. Incidents should be tracked and fines issued. The public should be educated about anti-dumping ordinances, trained to recognise illegal dumping, and asked to report it. The success of anti-dumping programs may depend on the availability, convenience, and costs of alternative disposal [2].

Past illegal dumping into storm drains may have caused accumulations of contaminated sediments, which now represent in-situ pollution and continue to release contaminants. Such sources need to be addressed and remediated usually by sediment removal prevention of future dumping [29].

Illicit connections to storm drains contribute to stormwater pollution [30, 31] and should be prevented through regulation, inspection, testing and education. Municipal building and plumbing codes must prohibit connections of sources other than stormwater to storm drains and establish penalties for violations. Building and plumbing inspectors enforce codes by visual inspections of sites during construction, and document their findings. Where needed, common methods of tracking connections are used for verifications (smoke, dye and TV testing). These programs may be costly and should be supported by public education activities [2]. Similar programs should be conducted in older areas [32]. These programs may be relatively costly, with requirements on equipment and well-trained teams employed in field testing. Public reporting of improper waste disposal/connections should be encouraged. These activities may be impeded by the lack of access to private property [2].

Finally, leaking sanitary sewers need to be also examined and corrected. Towards this end, dry weather infiltration and inflow, wet-weather overflows and leaking sanitary sewers are detected, using source identification techniques. Again, costs are significant and public reporting of problems is helpful. Limited access to private property impedes these activities [2].

3.4. HOUSEKEEPING PRACTICES

The availability of toxicants for entry to stormwater can be reduced by good housekeeping practices employed by the general public, municipal employees, small business and others. These measures focus on introducing and following good procedures for storage, handling and transport of materials, which could end up in stormwater. Successful implementation requires education and training.

With respect to storage, materials should be stored inside or under cover on impervious bases, with means of secondary containment. Inspections of storage and containers for damage and leaks should be conducted regularly. Non-roofed material storage should be designed to direct runoff away from the stormwater system. In general, storage and handling should be minimised. For storage of some chemicals, existing regulations must be followed. Similar recommendations apply to unloading and loading areas [4].

Vehicle spill control needs to be practised through maintenance and spill cleanup preparedness. Spills will always happen, but their frequency should be reduced and sources eliminated. Towards this end, recommended measures include performing fluid removal inside, or under cover on paved surfaces, proper storage of hazardous materials, keeping spill cleanup materials ready, immediate cleanup of spills by dry cleanup methods, and preparation of a contingency plan outlining responsibilities. Hazardous/toxic spills should be reported to the appropriate authorities [2, 4].

Furthermore, vehicles and service/parking areas should be inspected and cleaned regularly, parking spots designated to facilitate leak tracing, and procedures for repairing/cleaning leakage should be developed. For vehicle wash down, special areas (e.g., grassed areas) with appropriate runoff treatment should be provided. Signage prohibiting disposal of pollutants into drainage should be installed. Refueling areas should have concrete bases and spills be cleaned by dry methods. Again, suitable signage should be provided - do not top off; inspect for leakage, do not hose off. Vehicle maintenance should be performed indoors; if done outdoors, designate a special area, use drip trays and dry clean up, and drain fluids from scrapped vehicles [4].

For larger operations, spill control, cleanup and reporting are usually covered by regulations. Efforts need to be directed towards educating the public to keep vehicles in good shape, and impressing repair shop owners to always identify leaks and advise car owners. For large spills, private specialist companies may be needed [4].

Above ground tank spills should be reduced or prevented by installing safeguards against accidental releases and secondary containment equipment, conducting inspections, and training for cleanup [2].

The last measures discussed here are vegetation controls, by both mechanical and chemicals means (the latter were already discussed earlier). These methods are particularly of concern on steep slopes, in vegetated drains and creeks, and in areas next to catchbasins and detention/retention basins. Essential considerations include management of chemical use [25, 26], use of indigenous vegetation requiring low maintenance, maintenance of healthy vegetation (including mowing/mulching), and proper disposal of vegetation clippings [2]. Dumping of clippings should be prevented by ordinances; clipping composting is preferable to landfill disposal. The public should be educated about the safe use of herbicides or alternatives, proper disposal of clippings, and use of vegetation in erosion control/prevention.

3.5. REDUCTION OF STORMWATER POLLUTION BY CONSTRUCTION ACTIVITIES

Construction activities need to be planned and managed to minimise their impact on stormwater quality. The steps included in such controls include erosion control, sediment collection, site water control, equipment storage and maintenance, materials storage, and litter control. The first four measures are discussed here; the remaining ones are addressed in Sections 3.3 and 3.6.

Most municipalities with advanced pollution controls require submission of sediment and soil erosion control plans with applications for construction permits [25, 26]. Essential measures to be applied include minimisation of land clearance, planning of construction activities outside of the wet season, limited site access, provision of site services to minimise disturbance, installation of temporary drainage systems, application of soil protection measures (chemical stabilisation, mulching), keeping soil/material stockpiles away from runoff flows, and reestablishing disturbed areas as soon as possible [4].

Released sediment should be retained and collected at the site, as much as possible, by such sediment control devices as grass filter strips, sediment traps, and various sediment filters. In general, generation of dust, litter and debris should be minimised and the site should be regularly maintained [4].

Recognising that rainfall and runoff are the forces causing soil erosion and transport, it is important to implement water control at construction sites. Steps to be taken include preparing plans identifying drainage boundaries and flows, diverting water away from sites susceptible to erosion, preventing erosion by reducing flows/velocities entering bare soil areas, providing bridges for vehicle crossing of drainage channels, and ensuring regular site inspection, maintenance and sediment cleaning [4].

Other measures related to control of construction activities include minimising the number and size of stockpiles, locating stockpiles away from drainage routes, covering and protecting stockpiles from rain, instituting litter control, providing waste disposal bins, cleaning the site daily, disposing properly of scrap, and educating staff re needs for litter control [4]. Finally, construction activities also involve operation and maintenance of motorised equipment, which should be properly stored and maintained to minimise risks of pollution or spills, as discussed in Section 3.3.

Similar measures apply to building repairs - construction materials should be stored under cover and the site be cleaned daily by dry methods. When painting, avoid leakage/disposal into storm sewers, collect paint residue (dry), and clean equipment/tools away from drainage areas. Building drainage should be inspected annually, and various elements (drain inlets, spouts, down pipes and pipes) cleaned twice per year [4].

3.6. MAINTENANCE ACTIVITIES

Maintenance is important for controlling accumulations of polluting materials in the catchment and keeping the drainage system operating as designed.

Street cleaning reduces accumulations of pollutants before their washoff by stormwater during runoff. The effectiveness of street cleaning as a pollution control measure is uncertain, with some reports of inconclusive results [33], as shown in Fig. 3.

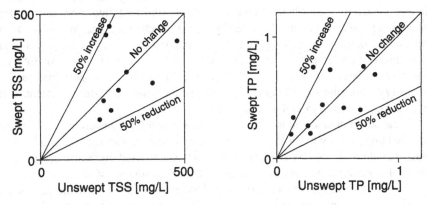

Figure 3. Bivariate plots of median EMCs (event-mean-concentrations) for swept and unswept conditions (after [33]).

Undoubtedly, effective street cleaning removes some sediments, debris and pollutants from street surfaces and improves street appearance. For best results, street cleaning should be done regularly, employing the best equipment in the most polluted areas. The cleaning frequency should be assessed with respect to inter-event times, and increased frequencies applied in pollution hot spots and/or just before the rainy season [5]. Street cleaning by flushing should be restricted and debris should be kept away from sewer inlets. Proper maintenance and operation of sweepers and good record keeping are important (i.e., km cleaned and tonnes of waste collected). Major limitation of street cleaning is caused by interference of parked vehicles and limitations of conventional equipment to handle fines and oil and grease. Public cooperation is needed with respect to parking and reduced littering [2, 25, 26].

Source controls should be also applied in maintenance of parks, conservation areas, golf courses and medians. The planning of such maintenance should start with determining the proximity of various source of pollution to water. Chemicals applied should be controlled by avoiding over-fertilising, minimising or avoiding the use of pesticides, removing litter and debris before grass mowing, reducing mowing in areas adjacent to water, keeping cuttings away from water, using mulch to absorb and filter water flows, and using indigenous plants to reduce water needs and improve infiltration [2, 4, 26].

Domestic waste collection and recycling collection help reduce littering and illegal disposal. Proper collection vehicles should be used to avoid spillage and dry methods should be used to collect spillage. Recycling containers should be closed. On-site cutting and green waste should be collected separately and composted. In public places, garbage receptacles should be provided, well identified and their good appearance maintained. Receptacles should be located near the source of litter, monitored and emptied frequently to avoid overflow (never more than 75% full) [2, 4, 26].

Catchbasins were introduced into drainage practice to intercept sediment and debris washed off from streets and thereby reduce sedimentation in sewers and sediment influx to wastewater treatment plants. More recently, concerns were raised about the impacts of catch basin contents washout (the first flush) on stormwater quality. These concerns can be alleviated by regular catch basin cleaning, which also prevents inlet clogging and restores the sediment trapping efficiency. Cleaning should be prioritised with emphasis on areas with high accumulation of pollutants. Recognising the large numbers of catch basins in urban areas, these operations can be rather costly and require special equipment and operator training. Decant contents can be toxic and may require special disposal [2, 4].

Some storm drains tend to clog with sediment and other materials, and require regular cleaning, usually by flushing. After locating sewers with deposit problems, a cleaning schedule should be developed and waste collected by vacuuming. For flushing, the equipment/facilities required include a source of water (hydrant), sediment collector (vacuum truck), inflatable bladders to block downstream pipe ends, and a sediment/containment/treatment facility if not discharging into sewers. Flushing into sanitary sewers, or detention areas (where feasible), is much less expensive. Well-trained field crews are required. Flushing is usually effective in small pipes; in other cases, it may be unfeasible [2].

Good maintenance of roadways and bridges also contributes to stormwater quality. Road pavements should be kept in good condition, by repairing cracks which are potential sources of pollutants. Repair materials should be kept away from drainage channels, which may require to block inlets during construction. Road and bridge drainage should be designed to collect and convey stormwater, and prevent inflow of runoff from adjacent areas. Special drainage facilities, including stormwater management and spill containment, may be included in specific cases [2].

Detention and infiltration facilities also require preventive and corrective maintenance to preserve their effectiveness. Typical operations include removal of silt and sediment from these facilities and their pretreatment devices, trash removal every 6 months, maintenance of vegetation (height < 0.4 m), and checking for problems. The associated costs may be high, particularly for removal and disposal of sediments. Depending on contamination of such sediments, special disposal procedures may have to be followed and may require contracting out. The public may play a useful role by reporting problems at these facilities [2].

It is important to maintain urban storm channels and creeks by preserving their conveyance capacity and avoiding sediment, waste and pollutant accumulations, which lead to other problems (pollution, habitat destruction, visual impairment, odours). The main task is regular removal of illegally dumped objects and materials from these channels. Towards this end, it is required to identify dumping hot spots, conduct inspections, post 'no littering' signs with phone numbers to call re dumping in progress, adopt and enforce substantial penalties, modify channels to improve hydraulics, maintain logs to evaluate materials removed, and establish buffer zones along creeks. Associated costs include program administration, purchase/installation of signs, costs of haulage of removed materials, rental of equipment to remove large items, landfill fees for disposing illegally dumped material, and cost of the equipment used in these operations. The public needs to be advised about the need for proper disposal of refuse and the applicable penalties. Promotion of volunteer collection of litter and clean out of creek is important. Program limitations - cleaning out may disturb aquatic species, access to private property is limited, tradeoffs need to be made between hydraulics and habitat conditions, and the issue of public safety [2, 4].

4. Concluding Remarks

For all stormwater management measures, some literature is available on their design, performance and maintenance [2,6,7,24]; however, practical experience with these measures is often missing, or relates to different climates. The selection of BMPs is empirically based, generally starting with application of source controls (policies) followed up by "structural" BMPs. The selection process starts with establishing the performance goals, listing solution alternatives, eliminating unfeasible measures, ranking the remaining measures with respect to benefit/cost ratios, and finally selecting the most effective combination of BMPs [24]. A recommended treatment train - the first element are SCs, then minimise connected impervious areas, at individual lots - swales, filter strips, infiltration trenches (Roesner and Urbonas, in Maidment). Source control measures should be selected with respect to their ability to meet regulatory requirements, effectiveness of practice to remove pollutants of concern, public acceptance of the practice, ability to implement, institutional constraints and costs. Such a selection process can be aided by ranking the applicable practices.

5. References

1. Marsalek, J. (1998) Challenges in urban drainage, in J. Marsalek, C. Maksimovic, E. Zeman and R. Price (eds), *Hydroinformatics tools for planning, design, operation and rehabilitation of sewer systems*, NATO ASI Series, 2. Environment, Vol. 44, Kluwer Academic Publishers, Dordrecht/Boston/London, pp. 1-23.
2. Water Environment Federation (WEF) and American Society of Civil Engineers (ASCE). (1998) Urban runoff quality management. WEF Manual of Practice No. 23, ASCE Manual and Report on Engineering Practice No. 87, WEF, Alexandria, VA, USA.
3. Urbonas, B.R. and Roesner, L.A. (1993) Hydrologic design for urban drainage and flood control, in D.R. Maidment (ed), *Handbook of Hydrology*, ISBN 0-07-039732-5, McGraw Hill, Inc., New York, pp. 28-1-28-52.
4. Environment Protection Authority (Victoria). (1998) Best practice environmental management guidelines for urban stormwater (draft). EPA, Melbourne, Australia.
5. Lawrence, A.I., Marsalek, J., Ellis, J.B.V., and Urbonas, B. (1996) Stormwater detention and BMPs. *Journal of Hydraulic Research* **34**(6), 799-813.
6. Ministry of Environment and Energy. (1994) Stormwater management practices planning and design manual. Ontario Ministry of the Environment, Toronto.
7. Schueler, T.R. (1987). Controlling Urban Runoff: A Practical Manual for Planning and Designing Urban BMPs. Washington Metropolitan Water Resources Planning Board, Washington, DC.
8. Schueler, T.R., Kumble, P.A., and Heraty, M.A. (1992) A current assessment of urban best management practices, techniques for reducing non-point source pollution in the coastal zone. Metropolitan Council of Governments, Washington, DC.

14

9. Yura, T., Takahashi, Y., Suzuki, K., and Jinbo, H. (1999) Plan to introduce stormwater infiltration facility and sustainable maintenance system with accompanying large scale housing site, in I.B. Joliffe and J. E. Ball (eds), Proc. of the 8th Conf. on Urban Storm Drainage, Sydney, Aug. 30-Sep. 3, pp. 990-997.

10. Coombes, P.J., Kuczera, G., Argue, J.R., Cosgrove, F., Arthur, D., Bridgeman, H.A., and Enright, K. (1999) Design, monitoring, and performance of the waters sensitive urban redevelopment at Figtree Place in Newcastle, in I.B. Joliffe and J. E. Ball (eds), Proc. of the 8th Conf. on Urban Storm Drainage, Sydney, Aug. 30-Sep. 3, pp. 1319-1326.

11. Zaizen, M., Urakawa, T., Matsumoto, Y., and Takai, H. (1999) Inspection of rainwater utilization at dome stadiums in Japan, in I.B. Joliffe and J. E. Ball (eds), Proc. of the 8th Conf. on Urban Storm Drainage, Sydney, Aug. 30-Sep. 3, pp. 1358-1365.

12. Sieker, F. (1998) On-site stormwater management as an alternative to conventional sewer systems: a new concept spreading in Germany, in Water Quality International 1998, Book 8 - Diffuse pollution, urban drainage, submarine outfalls, pp. 66-71.

13. Urbonas, B. (1994) Assessment of stormwater BMPs and their technology. *Water Science and Technology* **29(1-2)**, 347-353.

14. Azzout, Y., Barraud, S., Cres, F.N., and Alfakih, E. (1994) Techniques alternatives en assainissement pluvial. Lavosier Tec and Doc, Paris, ISBN: 2-85206-998-9.

15. Pratt, C.J., Newman, A.P., and Bond, P.C. (1998) Mineral oil bio-degradation within a permeable pavements: long term observations, in Proc. Of NOVATECH Conf., Lyon, France, May 4-6, 1998, pp. 501-507.

16. Hamacher, R. and Haussmann, R. (1999) Infiltration of stormwater: a concept for urban drainage planning; construction and maintenance costs, in I.B. Joliffe and J. E. Ball (eds), Proc. of the 8th Conf. on Urban Storm Drainage, Sydney, Aug. 30-Sep. 3, pp. 1217-1224.

17. Greb, S. R., Corsi, S., and Waschbusch, R. (1998) Evaluation of Stormceptor® and Multi-Chamber Treatment Train as urban retrofit strategies, in Proc. National Conference on Retrofit Opportunities for Water Resources Protection in Urban Environments, Chicago, IL, Feb. 9-12, 1998, pp. 277-283.

18. Pitt, R., Robertson, B., Barron, P., Ayyoubi, A., and Clark, S. (1999) Stormwater treatment at critical areas: The Multi-Chamber Treatment Train (MCTT). U.S. EPA, report EPA/600/R-99/017, Cincinnati, Ohio.

19. Urbonas, B. (1999) Design of a sand filter for stormwater quality enhancement. *Water Environment Research* **71**, 102-113.

20. Mothersill, C.L., Anderson, B.C., Watt, W.E., and Marsalek, J. (2000) Biological filtration of stormwater: field operations and maintenance experiences. *Water Qual. Res. J. Canada* **35(3)**, 541-562.

21. Lau, Y.L., Marsalek, J., and Rochfort, Q. (2000) Use of a biofilter for treatment of heavy metals in highway runoff. *Water Qual. Res. J. Canada* **35(3)**, 563-580.

22. Raimbault, G., Nadji, D., and Gauthier, C. (1999) Stormwater infilrtation and porous material clogging, in I.B. Joliffe and J. E. Ball (eds), Proc. of the 8th Conf. on Urban Storm Drainage, Sydney, Aug. 30-Sep. 3, pp. 1016-1024.

23. Makepeace, D.K., Smith, D.W., and Stanley, S.J. (1995) Urban stormwater quality: summary of contaminant data. *Crit. Rev. Environ. Sci. Technol.* **25:** 93-139.

24. Camp Dresser McKee, Larry Walker & Associates, Uribe & Associates, and Resource Planning Associates. (1993) California stormwater best management practice handbooks.

25. Novotny, V. and Olem, H. (1994) *Water quality - prevention, identification, and management of diffuse pollution.* Van Nostrand Reinhold, New York, ISBN 0-442-00559-8.

26. Novotny, V. (ed). (1995) *Nonpoint pollution and urban stormwater management.* Technomic Publishing Inc., ISBN 1-56676-305-3, Lancaster (USA)-Basel.

27. Lord, B.N. (1988) Program to reduce deicing chemical usage, in L.A. Roesner, B. Urbonas and M.B. Sonnen (eds), Design of urban runoff quality controls, ISBN 0-87262-695-4, ASCE, New York.

28. Environment Canada and Health Canada. (2000) Priority Substances List Assessment report: road salts. Ottawa, August.

29. Hubbard, T.P. and Sample, T.E. (1989) Source tracking of toxicants in storm drains. In: Proc. Design of urban runoff quality controls, Engineering Foundation Conference, published by ASCE, New York, pp. 436-447.

30. Schmidt, S.D. and Spencer, D.R. (1986) The magnitude of improper waste discharges in an urban stormwater system. *JWPCF* **58:** 339-348.

31. U.S. Environmental Protection Agency (EPA). (1993) Investigation of inappropriate pollutant entries into storm drainage systems. A user's guide. Report EPA/600/R-92/238, U.S. Environmental Protection Agency, Office of Research and Development, Cincinnati, Ohio.

32. Johnson, B., Tuomari, D., and Sinha, R. (1998) Impacts of on-site sewage systems and illicit discharges on the Rouge River, in Proc. National Conference on Retrofit Opportunities for Water Resources Protection in Urban Environments, Chicago, IL, Feb. 9-12, 1998, pp. 132-135.

33. U.S. Environmental Protection Agency (EPA). (1983) Results of the Nationwide Urban Runoff Program Volume I - Final Report. Water Planning Division, U.S.EPA, Washington, DC.

REVIEW OF STORMWATER SOURCE CONTROLS IN AGRICULTURAL CATCHMENTS

W.E. WATT
Queen's University, Kingston, ON, Canada K7L 3N6

1. Introduction

Stormwater runoff from both melt and storm rainfall on agricultural lands can result in a variety of impacts: flooding, ecosystem degradation, groundwater contamination, and degradation of surface water quality in a similar manner to urban stormwater as described in a companion paper [1]. The agricultural and urban cases are similar in that both agricultural development and urban development affect the land phase of the hydrologic cycle. Calder [2] lists intensification of agriculture as one of four major land-use changes (the others are afforestation & deforestation, draining of wetlands, and urbanization) in terms of hydrological effects. He notes that "intensification of agriculture involving land drainage, the use of fertilisers and pesticides, and the stall and battery farming of animals and poultry represents another major land use change which may have impacts on soil erosion and water quality and quantity." In both rural and urban cases, there are impacts at both the lot scale and the watershed scale; however, in the agricultural case, the lot is either the farmstead, a portion of the farm or the entire farm.

Calder [2] classifies the major water quantity and quality impacts/concern of agricultural intensification as hydrologic effects, nitrate concentrations, pesticides, farm wastes, and soil erosion. In some jurisdictions, habitat destruction would be added to this list, which is further discussed below.

- **Hydrologic Effects** include changes in the quantity and the timing of runoff.
- **Nitrate Concentrations** in the surface waters of agricultural catchments have been rising steadily, largely due to of heavy application of inorganic fertilisers.
- **Pesticides,** which can be toxic to humans and wildlife, do not always break down before reaching a watercourse and can be subject to aerial transport, which results in their widespread and continued occurrence in many parts of the world.
- **Farm Wastes,** including manure, are all potential sources of pollution for both surface and groundwater, and incidents of pollution are increasing as livestock operations have intensified.
- **Soil Erosion** as a result of agricultural activities is a worldwide problem. Hydrologic consequences include siltation of watercourses and reservoirs, increased flood peaks and a reduction of low flows as a result of decreased vegetation and soil.

17

J. Marsalek et al. (eds.), Advances in Urban Stormwater and Agricultural Runoff Source Controls, 17–26.
© 2001 *Kluwer Academic Publishers. Printed in the Netherlands.*

- **Habitat Impacts/Concerns**, result, in part, from sediments, nutrients and pesticides mentioned above, and from clearing riparian vegetation (trees and shrubs) along shorelines, land clearing, drainage ditch construction, and natural channel straightening [3].

Another difference between the urban and agricultural cases is that typically, in the agricultural case, there is no municipality to impose regulations. At the province or state level, regulations are usually specified by an environment agency (occasionally a fisheries agency) and apply to water quality (not water quantity). Guidelines are usually prepared by an agriculture agency and address both water quantity and water quality issues. Another difference is the stronger focus on non-structural measures (as opposed to structural) in the agricultural case.

The terminology on source controls is less well-defined for the agricultural case than for the urban case. In fact, neither the word source nor the word controls appears to be widely used by agricultural science researchers and practitioners. The term management practices is more common than controls, but any classification involving distance from the source does not appear to be common, as it is, for example in the urban case. Accordingly, in this paper, the definition adopted by Marsalek [1] for the urban case will be adopted for the rural case as well, that is "measures designed to control the generation of, and entry of pollutants into, stormwater runoff, with emphasis on non-structural and semi-structural measures applied at or near source." However, it should be noted that the source is typically larger in area than in the urban case. Even the smallest field is usually larger than the largest parking lot!

The primary objective of the review was to establish a framework for discussions of source controls in agricultural catchments at the NATO Advanced Research Workshop on Source Control Measures for Stormwater Runoff which was held in St. Marienthal, Germany, in November, 2000. The review starts with introduction to the problem and includes a discussion of source control measures for water quantity (which also affect water quality), and a discussion of source control measures for enhancing water quality.

2. Source Control Measures Reducing Stormwater Runoff

Source control measures for reducing stormwater runoff are considered at two levels: the farmstead and the field. At the farmstead level, they are essentially the same as those in urban areas and include minimising the extent of connected impervious areas and diverting runoff from impervious to pervious areas. These measures are intended to control stormwater quantity but they also improve surface water quality by diverting pollutants to soils. Care must be taken to prevent the entry of pollutants to groundwater. At the field level, the measures are quite different than the urban case; they include conservation tillage, cover cropping, and maintaining/restoring wetlands. A brief overview of selected measures follows.

2.1 CONSERVATION TILLAGE

2.1.1 *Conservation Tillage to Reduce Runoff and Erosion*

Conservation tillage has been defined by the Soil Conservation Society of America as a form of non-inversion tillage which maintains protective amounts of crop residue on the soil surface after planting a crop [4]. It includes various systems ranging from chiselling to no till (also called direct seeding), in all cases intended to reduce runoff and soil loss through erosion, and more recently pesticide losses by surface runoff. Conservation tillage was widely used in the southern U.S. as early as 1984 [5]. Moore and Larson [6] noted that infiltration, surface water storage, and erosion have all been shown to be directly affected by tillage and residue. There exists a body of literature on the impacts of various systems on runoff volume, peak flow rate, erosion rates, phosphorus losses and pesticide losses. A few examples are given below.

- Andraski et al. [7] reported that peak flow rates were reduced by conservation tillage (compared to conventional tillage) under natural and simulated rainfall.
- Yoo and Rochester [8] point out that peak flow rates are as important as runoff volumes in causing overland flow erosion. They studied tillage systems of cotton for 40 storm events over three years and reported that reduced tillage with a winter wheat cover crop was the most effective in reducing runoff volume and peak flow rate.
- Blough et al. [9] compared four tillage systems (conventional, chisel and slit, all with disking, and slit without disking) on a silt loam under simulated rainfall. They found that the infiltration and surface storage created through slit tillage nearly eliminated surface runoff until the slit overflowed, in this case by the runoff from the 2-year storm. In addition, runoff and erosion at the end of the 40-year, 30-minute storm (90 mm/h) were significantly larger for conventional tillage with no significant difference among the other three systems.
- van Vliet *et al.* [10] compared the effects of three tillage treatments (conventional, reduced and no tillage) on seasonal runoff and soil loss in the Peace River region of Canada. They found reduced annual erosion with no tillage (50% reduction) and reduced tillage (20% reduction), and that eliminating cultivation in the fall protects the soil from snowmelt erosion.
- Schmidt and Zimmerling [11] report on a study using simulated rainfall in the Saxonian part of the Neisse watershed in Germany on cropped soils with different tillage systems.

2.1.2 *Tillage Tradeoffs*

Although conservation tillage practices generally reduce runoff and soil loss through erosion, they can result in degraded water quality of both groundwater and surface water [12]. First, consider groundwater. Infiltration is increased under conservation tillage, and this increased soil water increases the possibility that nitrates and pesticides will enter tile drainage water or groundwater. Second, consider surface water. The amount of organic matter in the soil is generally increased under no till or reduced till practices and, as it decays, more soluble phosphorus becomes available for runoff. In addition, carbon, nutrients, and pesticides readily attach to the fine

20

particles, which are more easily eroded than the coarse particles. This increase in concentrations may offset the reduction in runoff volume such that loadings increase.

2.2 COVER CROPPING

Cover crops provide a protective cover that reduces the risk of runoff and erosion between growing seasons. They include intercrops that are grown along with the main crop and green manures that are generally sown after the main crop has been harvested [12]. A secondary benefit of cover crops is that they act as a sink for nutrients remaining in the soils. A few Canadian examples of cover crop studies are given below.

- In Ontario, intercropping silage corn with red clover reduced runoff by 40-87% [13].
- In Quebec, intercropping grain corn with timothy-alfalfa mix reduced soil loss by 35% and phosphorus loss by 25% [4].
- In Prince Edward Island, winter wheat and straw mulching after potato crops reduced nitrate levels in the drainage water by 30-50% [14].

2.3 MAINTAINING AND RESTORING WETLANDS

The National Wetlands Working Group [15] define a wetland as

"... land that has the water table at, near, or above the land surface or which is saturated for a long enough period to promote wetland or aquatic processes as indicated by hydric soils, hydrophytic vegetation, and various kinds of biological activity that are adapted to the wet environment."

2.3.1 Natural Wetlands

Canada, which has more than 14% of its area as wetlands, accounts for about 24% of the world's wetlands. In southern Canada, more than half of the original wetlands have been drained, about 85% as a result of agriculture. In recent years, this drainage trend has been somewhat reduced as a result of an increased environmental consciousness in general and a better understanding of the benefits of wetlands, participation by waterfowl and wildlife organizations, and government policies and programs.

- Environmental benefits of wetlands include providing habitat for wildlife, improving water quality, recharging groundwater, augmenting low flows and attenuating flood flows.
- Larger wetlands can also provide recreational, educational and economic benefits.
- Waterfowl such as ducks and geese depend on wetlands and are affected by poor water quality.
- The North American Waterfowl Management Plan seeks to restore waterfowl populations in North America to levels recorded during the 1970s. In 13 years of operation, more than 13 million ha of wetland ecosystem have been conserved [16].

2.3.2 *Constructed Wetlands*

Constructed wetlands are being used on farms in eastern Canada to treat manure runoff and milkhouse waste. Removals in terms of fecal coliform counts as high as 99 percent have been realised [3]. In the Atlantic Provinces, several artificial wetlands are used as tertiary treatment of domestic wastewater and agricultural runoff from manure storage and feedlots and to create habitat [3].

3. Source Controls Enhancing Stormwater Quality

Source controls for enhancing stormwater quality are policies and practices that are intended to reduce or eliminate the entry of pollutants into runoff from agricultural areas. Bernard et al. [12] point out that "Farmers can help to improve water quality in three main ways: controlling the processes that move soil and agricultural inputs into water (e.g., erosion, runoff, and drainage), improving the way in which agricultural inputs (e.g., fertilisers, manure, and pesticides) and waste are managed, and making better use of buffer zones and shelter belts." A brief discussion of selected measures follows.

3.1 EDUCATION, AWARENESS AND PARTICIPATION

Education, awareness and participation by the agricultural industry, local communities, and government are essential for planning implementation of source control measures to prevent water quality degradation.

3.1.1 *Role of the Agricultural Industry*

The agricultural industry can create and promote codes of practice and peer advisory programs, and encourage farmers to develop environmental farm plans [12].

- Codes of practice comprise guidelines that producers can follow to ensure their management practices are environmentally stable. In some jurisdictions (e.g., the Province of British Columbia), codes of practice are embedded in legislation.
- Peer advisory programs help farmers understand environmental sustainability and avoid penalties. In British Columbia, producer organisations operate peer advisory services to help resolve nuisance and pollution complaints against farms. The peer advisor investigates complaints, advises on necessary corrective measures, conducts a follow-up visit, and refers the case to the appropriate government agency if the problem persists.
- Environmental farm planning is a way to involve farmers and help make their operations more environmentally sustainable. This plan is prepared voluntarily by a farm family to identify environmental strengths and weaknesses, and to set realistic goals to improve environmental conditions. Ontario's Environmental Farm Plan Program began in 1993.

3.1.2 *The Role of the Community*

Experience in Canada and the United States is that programs to improve water quality in rural watersheds can succeed only through the active participation of the rural community [12]. The concerned community must agree on the need for action to improve water quality in the target watershed. Usually, the diversity of interests require a coalition including all stakeholders in the watershed. Canadian examples include the following:

- the Boyer River in Quebec, a watershed of 21,700 ha, of which 60% is farmland, much in high-density livestock production (mostly hogs), where nutrient laden runoff from the farms has contributed to pollution of the river;
- the Grand River in Ontario, a watershed of 7,000 km² where 70% of the total phosphorus loading is from rural non-point sources; and
- the Oldman River in Alberta, a watershed of 17,000 km² (at Lethbridge) and the most intensive agricultural area in Alberta featuring irrigated agriculture and intensive livestock operations [12].

3.1.3 *The Role of Government*

Governments can provide education and training, enact legislation and create regulations, policies and programs, and develop incentive mechanisms.

- Policies and programs have been developed to promote activities that prevent pollution and promote the adoption of good management practices.
- Incentive mechanisms can be negative or positive. Negative incentives include taxes on farm inputs, such as fertiliser and pesticides. Positive incentives can be in the form of grants to farmers for adopting an environmentally sound practice or payment for adoptions of environmentally benign products.

3.2 PROCESS CONTROL/LAND MANAGEMENT

Process control involves controlling the processes (erosion, runoff, drainage) that move soil and agricultural inputs such as fertilisers, manure, and pesticides into the water [12].

Most of the practices employed to minimize water quality degradation were developed originally to control runoff and soil erosion. Of these, conservation tillage and cover crops have been discussed in section 2 of this paper. An additional practice, not previously discussed, is crop rotations. Crop rotations that include forages such as grass or clover improve the soil structure and reduce runoff, erosion and nitrate leaching [12].

3.3 INPUT AND WASTE MANAGEMENT

Input management includes integrated pest management practices and nutrient management practices so as to reduce the amount of these inputs available to be moved off farmland. "Waste management involves both reducing the amount and hazardous composition of waste, and undertaking its safe handling storage and dispersal" [12].

Integrated pest management involves first a combination of practices to reduce the amount of pesticides used and second, selection of, and control of the timing and method of application of, pesticides. OMAFRA [17] lists the following practices to reduce pesticide use: site selection, cultivar selection, nutrient and water management, crop rotations, planting and harvesting to avoid pest peaks, weed removal, sanitation, trap crops, and biological controls. Good pesticide practices include 1) selection of pesticides that are target specific, least persistent and low toxicity, low vapour pressure and low susceptibility to leaching, 2) adherence to guidelines for application rate and conditions, 3) selection of application methods that are target-specific rather than general, and 4) regular calibration of equipment.

Nutrient management involves first determining how much nitrogen, phosphorus, and potassium is in the soil and then 1) accounting for all nutrient sources, 2) fertilising according to a nutrient management plan, 3) using nutrients in a way that optimises uptake, and 4) preventing the buildup of nutrients in upper horizons of the soil.

To prevent contamination of groundwater and surface water, a management scheme for manure should involve handling, storage, and timing of land application. Manure is stored in either liquid form or solid form and in either open or closed storage systems. In Canada, the least environmentally safe method, open storages, were the most widely used methods in 1995 (e.g. 91% of cattle producers stored manure in solid form, mainly in an open pile with no roof). The safest method, covered storage, was seldom used, except on chicken farms where 45% used covered tanks [12]. The timing of manure application is also very important, in particular so as to avoid applications to frozen ground when runoff potential is highest. Surface runoff with a high concentration of ammonium may be toxic to fish in the receiving waters. Even when the ground is not frozen, manure should not be applied just before runoff-producing rainfall. In reporting the results of a plot test using simulated rainfall, Grando [18] reported much larger counts of total coliforms and fecal streptococci at shorter intervals (1 h) between application of liquid manure and commencement of rainfall than at longer intervals (24, 48 and 120 h). An additional consideration with respect to manure application is the impact of cover. van Vliet et al. [12] compared two manure treatments on corn land in south coastal British Columbia. They found a lower potential for nitrogen loading when manure was applied to a winter cover crop than when it was applied to exposed soil surface.

Sources of waste in agricultural operations include milkhouse wash water, livestock housing wash water, silo seepage, exercise yard and feedlot runoff, dead stock, and used oil and pesticide containers [12]. Many state and provincial agencies (e.g. [19]) have guidelines or manuals on best management practices to deal with these wastes.

3.4 BUFFER ZONES AND SHELTER BELTS

Buffer zones and shelter belts are intended to prevent contaminants leaving agricultural fields from entering waterways. Instead, they are trapped in buffer zones at the edges of fields or in locations where the runoff occurs [12]. Buffer zones are areas/bands of vegetation between agricultural land and water bodies. They are generally covered with grasses or other natural vegetation and vary in width. For example, in a series of environmental guidelines for various agricultural sectors [20], the Government of British Columbia specifies buffer zones of various width for controlling surface water runoff (e.g. berry producers, width = 10 m). Buffer zones improve water quality by filtering runoff, thereby reducing the loading of both dissolved and suspended sediments and other pollutants. Shelter belts are areas of trees and shrubs, either natural or planted, generally used to reduce wind erosion. They improve water quality in at least two ways: 1) by reducing atmospheric transport and deposition of fine soil particles and spray aerosols, and 2) by reducing the transport of snow and snowmelt runoff.

Studies on the performance of filter strips in terms of pollutant removal indicate a range of efficiencies for sediment, phosphorus and nitrogen/nitrates. Mickelson and Baker [21] report sediment-trapping efficiencies of 50-90% for both grass buffer strips and forested buffers. The results for phosphorus indicate removal efficiencies of 20-90% for total phosphorus and marginal removals for dissolved phosphorus. The removals for nitrogen/nitrates are similar to those for phosphorus with particulate forms removed more effectively than dissolved forms. Results of one study indicated removal efficiencies of 63-76% for total nitrogen and 27-57% for nitrates.

In recent report [22], the Ontario Soil, Water and Air Research and Services Committee state that "in Canada, research on the effectiveness of buffer strips in reducing non-point source pollution is virtually non-existent. Management practices should be developed for Ontario soil and climatic conditions. There are a number of questions that need to be answered:
1. What is the optimal crop species, width and management regime required to control non-point source pollution to a predetermined limit?
2. How should buffer strip width and type be varied with different soil textures and slope conditions?
3. How are buffer strips defined and managed for pastures adjacent to streams where the vegetation type does not change, and there is no tillage to contend with?

Additional research should be conducted to address these questions for sustainable crop production and improved environmental quality. Some of these basic questions pertaining to the width and type of vegetation needed to achieve a certain level on non-point source pollution control have never been investigated."

4. Concluding Remarks

(a) Urbanisation and intensive agricultural operations are similar in many ways and different in others in terms of both impact on aquatic ecosystems and practices (both structural and non-structural) to mitigate these impacts. Source controls are one such practice; they take different forms at the farmstead and field levels.

(b) In this paper, source controls are classified as those primarily directed at controlling runoff quantity (which also mitigate water quality impacts) and those concerned with runoff quality. As is the case in urban areas, the former generally have a longer history than the latter. This classification is somewhat artificial. For example, early practices for reducing runoff also reduced erosion and sediment in receiving waters. However, the focus on erosion was soil loss rather than quality of the receiving waters or ecosystem degradation so that the classification still applies.

(c) Measures for reducing runoff and erosion, which are applied at the field level, include conservation tillage, cover cropping, and maintaining/restoring wetlands. Measures for enhancing water quality, which are applied both at the farmstead and field levels, include education, awareness and participation, process control, input and waste management, and buffer zones and shelter belts.

(d) For each of the above measures, there are both general and specific guidelines, and demonstrated performance through research and demonstration projects. However, at the field level in particular, there is far more variation in acceptable measures than in the urban case. This is due primarily to the wider variation in the natural environment than in the built environment, but also to the wider variation in economic incentives/disincentives.

(e) Another difference between the urban and agricultural cases, in some instances, is the proximity of the source to the upper ends of the impacted aquatic ecosystems and the general affinity that farmers have to the land. As a result, one might expect more receptive audience for education, awareness and participation measures.

5. References

1. Marsalek, J. (2000) Review of stormwater source controls in urban drainage. Preprint, NATO ARW on Source Control Measures for Stormwater Runoff, St. Marienthal, Germany.
2. Calder, I.R. (1992) Hydrologic effects of land-use change, in D.R. Maidment (ed), Handbook of Hydrology, McGraw-Hill, Inc., New York, NY pp. 13.1-13.50.
3. Gregorich, L.J. (2000) Ecological issues, in D.R. Coote and L.J. Gregorich (eds), *The Health of Our Water*, Agriculture and Agri-Food Canada, Ottawa, ON, p. 77
4. Soil Conservation Society of America. (1982) *Resource Conservation Glossary*.3rd ed., Ankeny, IA.
5. Conservation Tillage Information Center. (1984) National survey of conservation tillage practices. Conservation Tillage Information Center, National Association of Conservation Districts, Ft. Wayne, IN.
6. Moore, I.D. and C.L. Larson (1979) Estimating micro-relief surface storage from point data. *Transactions of the ASAE* 21(1): 101-104, 109.

7. Andraski, B.J., T.C. Daniel, B. Lowery and D.H. Mueller (1985) Runoff results from natural and simulated rainfall for four tillage systems. *Transactions of the ASAE* 28(4): 1219-1225.
8. Yoo, K.H. and E.W. Rochester (1989) Variation of runoff characteristics under conservation tillage systems. *Transactions of the ASAE* 32(5): 1625-1630.
9. Blough, R.F., A.R. Jarrett, J.M. Hamlett and M.D. Shaw (1990) Runoff and erosion rates from slit, conventional, and chisel tillage under simulated rainfall. *Transactions of the ASAE* 33(5): 1557-1562.
10. van Vliet, L.J.P., R. Kline and J.W. Hall (1993) Effects of three tillage treatments on seasonal runoff and soil loss in the Peace River Region. *Can. J. Soil Sci.* 73: 469-480.
11. Schmidt, W. and B. Zimmerling (2000) Conservation tillage – an element of a stormwater control strategy. Preprint, NATO ARW on Source Control Measures for Stormwater Runoff, St. Marienthal, Germany.

26

12. Bernard, C., G.L. Fairchild, L.G. Gregorich, M.J. Goss, D.B. Harker, P. Lafrance, B. McConkey, J.A. MacLeod, T.W. Van der Gulik, L.J. P. van Vliet, and A. Weersink (2000) Protecting water quality, in D.R. Coote and L.J. Gregorich (eds), *The Health of our Water*, Agriculture and Agri-Food Canada, Ottawa, ON, pp. 99-110.

13. Wall, G.J., E.A. Pringle and R.W. Sheard (1991) Intercropping red clover with silage corn for soil erosion control. *Con. J. Soil Sci.* 71:137-145.

14. Milburn, P., J.A. MacLeod and S. Sanderson (1997) Control of fall nitrate leaching from early harvested potatoes on Prince Edward Island. *Can. Agric. Engr.* 39:263-271.

15. National Wetlands Working Group (1988) *Wetlands of Canada, ecological land classification series no. 24.* Sustainable Development Branch, Environment Canada, Ottawa, Ontario, and Polyscience Publications Inc., Montreal, Quebec. 452 p.

16. Cox, K.W. (1993) Wetlands: a celebration of life. Final Report of the Canadian Wetlands Conservation Task Force. North American Wetlands Conservation Council (Canada), Ottawa, ON.

17. OMAFRA (1999) Best management practices series: field crop production; fish and wildlife habitat management; horticultural crops; integrated pest management; irrigation management; livestock and poultry waste management; no-till; making it work; nutrient management; nutrient management planning; pesticide storage, handling, and application; soil management; water management, water wells. Ontario Ministry of Agriculture, Food and Rural Affairs, Guelph, ON.

18. Grando, S. (1996) Effets de deux modes d'epandage de lisier de porc sur la quanlité de l'eau de ruissellement. Memoire de Fin d'Etudes, ENITA de Bordeaux, France.

19. MOAFF (1992) Environmental guidelines for dairy producers 3: pollutants. British Columbia, Ministry of Agriculture, Fisheries and Food. Victoria, BC.

20. MOAFF (1994) Environmental guidelines for berry producers, 6: water management. British Columbia Ministry of Agriculture, Fisheries and Food, Victoria, BC.

21. Mickelson, S.K. and J.L. Baker (1993) Buffer strips for controlling herbicide runoff losses. Amer. Soc. Agric. Engr./Can. Soc. Agric. Engr. Meeting Paper No. 93-2084, Spokane, Wash.

22. OMAFRA (1998) Ontario soil, water and air research and services committee 1998 report to Ontario agricultural services coordinating committee. Ontario Ministry of Agriculture, Food and Rural Affairs, Toronto, ON.

OBJECTIVES OF STORMWATER MANAGEMENT - A GENERAL COMPARISON OF DIFFERENT MEASURES

H. SIEKER
Ingenieurgesellschaft Prof. Dr. Sieker mbH (IPS),
Berliner Str. 71, 15366 Dahlwitz-Hoppegarten, Germany

1. Introduction

Legislative bodies prescribe standards or target values for the design of environmental protection measures to achieve a certain level of quality, for example the water quality, or to preserve the water balance. As a basis for the definition of these quality levels overall environmental concepts are useful. These concepts describe an ideal situation, not a concrete target. However, they can be used to assess the actual situation. The overall environmental concepts have to be defined by the society as a whole not only by experts [1].

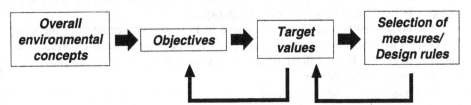

Figure 1: Translation of overall environmental concepts into practice [4]

Examples for such overall environmental concepts are:
- The demand for the protection of the environment in the German constitution
- The principles of sustainability (Brundtland report)
- Sections in the German Water Act (§1a) or the US Clean Water Act.

Because these overall environmental concepts do not consider technical possibilities and financial aspects, more realistic objectives or general goals are needed. These objectives are the basis of the definition of environmental standards. According to this
- The demand for a "good" water quality in the EU-water framework directive (not a very good quality is demanded) or
- The general goal of the US Clean Water Act that every water body should be "fishable and swimmable"

are objectives. For practical planning tasks these objectives have to be translated into concrete target values or standards. The target values are the basis for the selection

J. Marsalek et al. (eds.), Advances in Urban Stormwater and Agricultural Runoff Source Controls, 27–37.
© 2001 *Kluwer Academic Publishers. Printed in the Netherlands.*

and finally for the design of a certain measure. Examples for target values are the maximum permitted BOD concentration of 15 mg/l or other effluent standards at the outlet of a wastewater treatment plant (WWTP) in Germany. With this target value (emission standard) the engineer can make a decision for example for an activated sludge treatment plant with denitrification and P-removal and he can design the WWTP using a standard design rule. It should be noted that receiving water quality standards, for example a minimum oxygen concentration in a lake or a river, could also be a target value. Figure 1 shows how overall environmental concepts can be translated into practice (at least in theory). An important aspect of this scheme is the regular validation, if the defined target values and design rules are really sufficient to achieve the objectives.

2. Objectives of Stormwater Management

The main overall environmental concepts as a basis for the objectives of stormwater management have been already mentioned in the introduction. One more general goal, which is of importance for the management of stormwater runoff in Germany, can be found in the Water Act §1a(2):

Everybody is obliged to do everything that is necessary, to avoid pollution of water bodies, to use water economically with respect to the water balance, to preserve the water balance and to avoid an increase in runoff.

The underlined part had been introduced in 1996 as a consequence of the large flood events in the Rhine area in 1993 and 1995. To fulfill these overall environmental concepts, objectives in the following areas have to be defined:

- Protection against flooding
- Water quality (quality objectives)
- Water balance (quantity objectives)
- Cost minimisation.

Concerning the protection against flooding, which is the classical goal of hydrological planning, several target values exist. Traditionally, the conveyance of a design storm is the typical target value. More recently it is common to prove by simulation models that a certain flood frequency is not exceeded.

In the area of water quality and water quantity several deficiencies can be realised. For the treatment of stormwater runoff in separate systems, for example, there are generally no target values, which have to be fulfilled in Germany. So in practice stormwater runoff is usually not treated at all [5]. And if the authorities ask for a treatment they mostly prescribe a certain measure (e.g. sand traps). Similar deficiencies can be observed in the field of CSO treatment [2].

For the objective of preserving the water balance, which is part of the German Water Act, also no concrete target values exist, though it would be rather simple to introduce the corresponding target values into practice: the differences between the components of the water balance before and after the development, e.g. the construction of houses, should be calculated. Target values could be that these differences should not be more than "x" percent [4].

From the author's point of view, these missing target values have to be introduced, or if that is not acceptable, the authorities should admit it and reduce the objectives.

3. Consequences for the Planning Process

If new target values, especially for an unchanged water balance and the treatment of stormwater runoff are introduced into practice, the classical process of planning has to be changed as well. At the moment the planning strategy is usually linear. If a new paved area has to be drained, first a sewer system is designed which fulfills the target value of the protection against flooding. Then in the next step, a retention pond is designed to fulfill the target value of a maximum discharge rate. Finally, and only if necessary, treatment of the stormwater runoff is designed. With this linear planning approach, it is not possible to find an optimal solution regarding several target values of equal importance [4].

Instead, it is necessary to develop a more integral planning approach. Several different scenarios should be developed and an assessment concerning the different target values should take place. In other fields of engineering, e.g. traffic planning and also in some fields of hydrology, such multi-criteria assessments have been common for many years. That means for the engineer, that the main question is not any more how to design a certain measure. Of course, that will always be important, but an examination of the number of technical regulations in Germany shows that for almost every measure a design procedure is available. On the other hand there is no technical paper that offers assistance to the engineer for the decision process. The main question is: *which is the best measure under the given local conditions regarding the different target values? [5].* As an assistance for this task a catalogue of different stormwater management measures is introduced in the next section.

4. Catalogue of Stormwater Management Measures

Many different best management practices for stormwater runoff are known today. Infiltration techniques, stormwater utilisation, treatment measures in separate or combined sewer systems are only some examples for different categories. All the different measures have different impacts on the water cycle and the cycle of materials. The costs of implementation and maintainance can also differ a lot. As mentioned before, the problem for the engineer is usually not the design of a specific measure. It is rather difficult to decide for or against a certain measure. In [3] a catalogue with information about more than fifty different measures has been arranged, by offering for each measure

- A short description,
- The effects regarding the different objectives (water balance, behaviour for different rainfall events, removal of pollutants, etc.)
- The main influential factors (land demand, soil parameters, etc.),
- Costs (investment, maintenance, life cycle), and

- Legal aspects.

This catalogue can be used as a "toolbox" for preselection of measures for example in the planning process of a master plan. It gives an overview over the different alternatives and a rough idea about their effects.

Of course, it is not possible to investigate the actual effects of a certain scenario, for example on the water balance in a larger catchment, with this catalogue. For this, a simulation model is needed. Because of the complexity it is not possible to present the catalogue here in detail. Therefore the following section shows the main results as a general comparison of different measures.

5. A General Comparison of Different Measures

The following figures are showing the effect of different stormwater management measures on the water balance, the behaviour for typical design storms, the potentials for removing pollutants, and the costs. It has to be mentioned that in practice the different effects and costs can vary a lot. It is not the aim of this comparison to find the optimal solution for stormwater management. There is no optimal solution due to the dependencies of many different factors and the variety of objectives!

Table 1: Acknowledged measures

Measure	Short description
Storm sewer	Storm sewer designed for a design storm with a return period of 5 years
Open drainage trench	Open drainage trench, 3.0 m wide, 1.20 m deep, slope of embankment 1:1, width of catchment 100 m
Retention pond, open	Retention pond, open, earthen, sealed at the bottom, permanent water depth 0.25m, maximum depth 2m, slope of embankment 1:5, throttle rate 5 ℓ/s/hectare, designed for a return period of 5 years, specific volume 380 m³/hectare
Retention pond, concrete	Retention pond, concrete, closed, depth 2 m, throttle rate 20 ℓ/s/hectare, return period of 5 years, specific volume 270 m³/hectare
Permeable hard surface	Permeable hard surface, area of voids 20%, infiltration capacity 10^{-5} m/s, max. water depth 5 mm, overflow connected to a sewer system
Infiltration swale	Infiltration swale, depth 30 cm, slope of embankment 1:2.5, infiltration capacity 10^{-5} m/s, designed for a return period of 5 years, no overflow
Infiltration trench	Infiltration trench, no overflow, filled with gravel, volume of pores 30%,
Swale-trench-elements	Swale-trench-elements, not connected to a sewer system, dimension of swale: see above, infiltration capacity 10^{-6} m/s

Table 1: Acknowledged measures (continuation)

Measure	Short description
Swale-trench-system	Swale-trench-elements connected to a sewer system, dimensions of swale and trench as above, drain pipe 10 cm above the bottom, throttle rate 3.6 mm/h or 10 l/(s.hectare, infiltration capacity 10^{-6} m/s
Stormwater utilisation (household)	Tank with a 4 m³, volume utilisation of stormwater for toilets and washing machine, consumption (4 persons) 150 l/day, rate of utilisation approx. 66%.
Stormwater utilisation (irrigation)	Tank with a 4 m³, volume utilisation of stormwater for garden irrigation, consumption 80 l/day between May and September
Green roof	Roof greened with moss and sedum, thickness of vegetation layer 6 cm, thickness of drainage layer 6 cm, drainage rate 100 l/(s. hectare)
Roof garden	Roof with a soil layer of 20 cm, thickness of drainage layer 10 cm, max. water depth 6 cm, drainage rate 100 l/(s. hectare)
Stormwater settling tank	Stormwater settling tank, design criteria (ATV-suggestion) 10 m/h, specific volume 10.8 m³/hectare,
Combined sewer overflow	Combined sewer overflow (CSO), critical discharge 15 l/(s hectare)
CSO-Tank	Combined sewer overflow tank, throttle rate 0,7 l/(s. hectare), specific volume 25 m³/hectare,
Screen	Screen (fine rack), width of openings 4 mm
Soil filter retention pond	Soil filter retention pond for treatment in combined or separate systems, specific volume 80 m³/hectare, max. depth 1m, infiltration capacity 10^{-4} m/s
Primary treatment at WWTP	Primary treatment at a waste water treatment plant, mean hydraulic residence time of 1.5 hours
Activated sludge treatment	Activated sludge treatment (100,000 People) biological stage and clarifier
Stormwater treatment WWTP	Extension of an existing WWTP capacity from 2 times to 4 times the dry weather flow

32

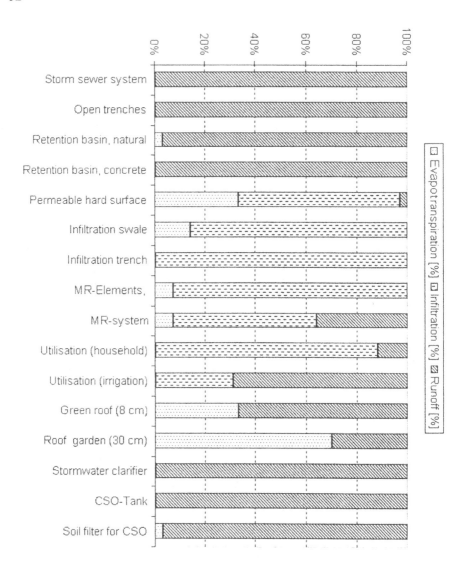

Figure 2: Mean annual water balance for different stormwater management measures

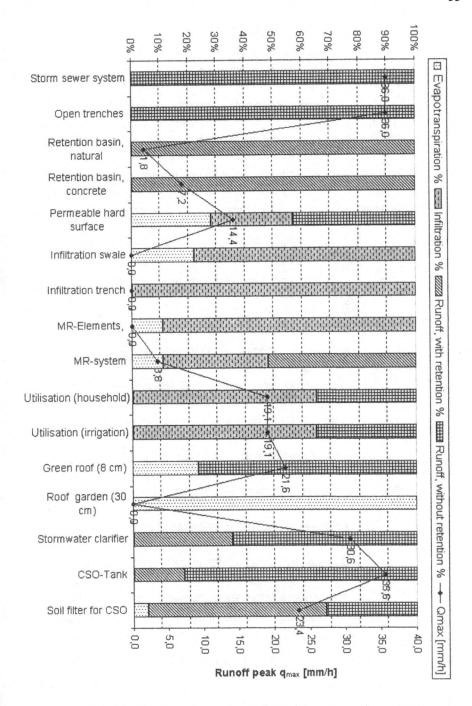

Figure 3: Water balance of different stormwater management measures for a typical design storm

34

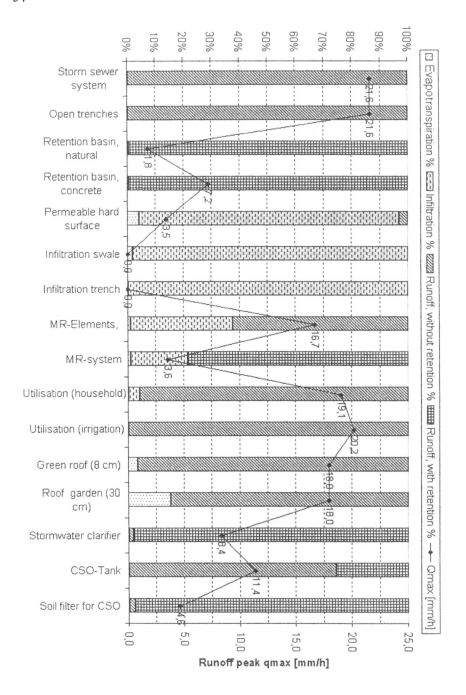

Figure 4: Water balance of different stormwater management measures for a rainfall period that caused a flood event (approx. 300 mm in 3 weeks)

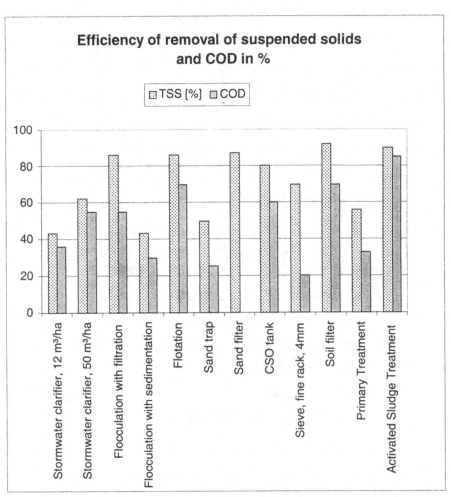

Figure 5: Removal of suspended solids and COD by different stormwater management measures

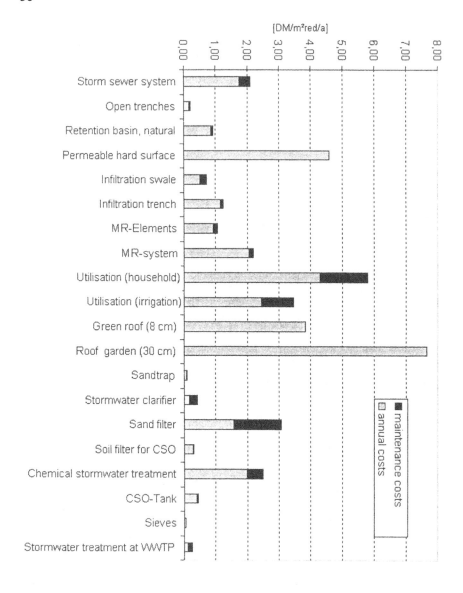

Figure 6: Total costs (annual capital costs plus maintenance costs) of different stormwater management
measures

measures

6. Summary

Undoubtedly, stormwater management has to consider several objectives derived from overall environmental concepts. For a translation of these objectives into practical measures, concrete target values are necessary. Objectives and target values are needed for flood protection, water quality and the water balance.

On the other hand, a large variety of different measures is available. These measures have different effects on different objectives and depend on many influential factors. The main future task for the engineer will be to find the "best" measure regarding the local conditions and the objectives.

To provide assistance for this task, a catalogue with information about different measures is presented. The measures are compared regarding the water balance, the behaviour for typical design storms, the potentials for removing pollutants and the costs. With this catalogue a preselection of measures, for example for a master plan, is possible.

7. References

1. Esser, B. (1997) Leitbilder für Fließgewässer als Orientierungshilfen bei wasserwirtschaftlichen Planungen, Wasser & Boden, 49. Jahrgang, Heft 4/99.
2. Sieker, H., Klein, M. (1998) Best management practices for stormwater runoff with alternative methods in a large urban catchment in Berlin, Germany, Water Science & Technology **38**, 91-97.
3. Mehler, R., Ostrowski, M. (1998) Integrating Better Stormwater Management Practices in Urban Stormwater Pollution Modeling, in: Modeling the Management of Stormwater Impacts, Volume 6, CHI Publications Guelph, Ontario, Canada.
4. Sieker, H. (2000) Generelle Planung der Regenwasserbewirtschaftung in Siedlungsgebieten, dem Fachbereich Bauingenieurwesen der Technischen Universität Darmstadt vorgelegte Dissertation (in press).
5. Sieker, H. (1999) Geographical Information Systems used as a Planning-Tool for On-Site Stormwater Management Measures, in J. Marsalek, W.E. Watt, E. Zeman and F. Sieker (eds), Proceedings of a NATO-Advanced Research Workshop Flood Issues in Contemporary Water Management, NATO Science Series Vol. 71, Kluwer Academic Publishers, Dordrecht/Boston/London, pp. 311-321.

BALANCING FLOOD CONTROL AND ECOLOGICAL PRESERVATION/RESTORATION OF URBAN WATERSHEDS

V. NOVOTNY, D. CLARK, R. GRIFFIN & A. BARTOŠOVÁ
Institute for Urban Environmental Risk Management
Marquette University, Milwaukee, WI 53201-1881, USA

1. Introduction

Throughout recent history, urbanisation has altered the ecological structure of urban streams and rivers. They have been channelised, constricted, and ultimately covered to gain space for urban development and to accommodate increased flood flows. The Los Angeles River in California and the Kinnikinnic River in Milwaukee represent the ultimate transformation of an urban stream into a concrete, high flow velocity channel with very little biological habitat. Concurrently, as a result of these modifications and diminished water quality, indigenous aquatic species have disappeared, resulting in a complete loss of sensitive species and the propagation of a few unwanted species tolerant to pollution and low quality habitat.

Urbanisation typically has an irreversible impact on natural drainage patterns and flows in the receiving water bodies impacted by development. If the development progresses in a planned, ecologically conscious way, the adverse impacts on population and properties can be minimal or minimised. Uncontrolled developments or past development in the flood plain that did not consider the impacts on hydrology, flood plain encroachment, morphology and ecology of the receiving water body system have had detrimental effects on the receiving water body, flood plain development and downstream uses of the water body.

Today, management of smaller and medium size urban streams must consider several objectives including

* Flood control
* Preservation and restoration of the ecological integrity of the receiving water body affected by point and nonpoint discharges and changes in hydrology and hydraulics
* Providing contact and noncontact recreation to the urban population
* Wastewater disposal and conveyance of polluted urban runoff
* Other uses such as water supply, navigation, or hydropower production

Some of the uses of the urban receiving water bodies conflict whereas others are complementary. For example, preservation and restoration of the ecological integrity and providing habitat for aquatic life complements recreational objectives. Indeed, a

J. Marsalek et al. (eds.), Advances in Urban Stormwater and Agricultural Runoff Source Controls, 39–56.
© 2001 *Kluwer Academic Publishers. Printed in the Netherlands.*

healthy stream ecology is a necessary condition for contact recreational uses. On the other hand, flood control often is in conflict with ecological and recreational objectives. In the context of watershed and water body management policy, these conflicts must be reconciled if an optimal policy is to be pursued.

In the past, urban engineers tried to resolve the problem of increased floods by increasing the flow capacity of urban streams. Common techniques included channel lining, covering and straightening of the stream. Such approaches are *conveyance-oriented*, i.e., the capacity of water bodies was increased by increasing the velocity in the channels. Conveyance-oriented flood control measures did not improve water quality, and they were detrimental to the habitat. Moreover, they passed flood control problems downstream. At the same time, development has continued to encroach on floodplains, exacerbating flooding problems. Traditional cost-benefit evaluations have often revealed negative net-benefits as costs often far exceed the benefits of flood damage reduction. The authors contend that past approaches to measuring benefits associated with flood control projects are incomplete for several reasons. First, they are based on the false premise that the only benefits are to those who experience flood damage. Second, they fail to fully characterise some of the ecological benefits that may be derived from some flood control projects.

Furthermore, flood control projects today are limited by the *antidegradation* clause of the water quality standards and control regulations. This rule specifies that a downgrading of good quality water bodies to something lower than good quality can occur, but not to a point of violation of water quality standards. Moreover, this can only occur if overriding regional socioeconomic effects are at stake. No downgrading of water quality is permissible for high quality water bodies. This rule makes implementation of past conveyance-oriented flood control approaches very difficult if not impossible. Today, most flood control projects must have a water quality improvement component. Consequently, *storage-oriented* approaches are preferred.

2. Hydrologic Changes by Urbanisation

Figure 1 shows a flood-frequency chart for Oak Creek, which is located in the Milwaukee (Wisconsin) metropolitan area that is undergoing rapid urbanisation. The area of the watershed is 69.8 km^2; presently the watershed is 44.6 % urbanised and the resident population is about 40,000. Cropland is the other dominant land use; however, the watershed is being rapidly urbanised. As Figure 1 illustrates, the flow that was a 100-year flood in the 1965 pre-development period could become a 2-year flood when the watershed is fully developed. The flow frequency curves were estimated from long term observations at the US Geological Survey gauging station. The annual daily flow maxima were corrected for the impact of urbanisation by a method proposed by McCuen [1].

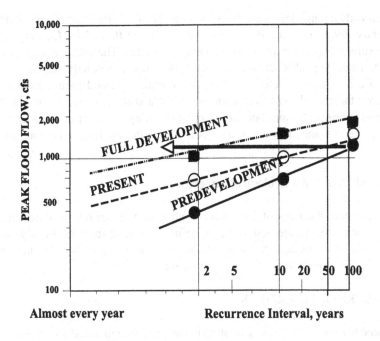

Figure 1. Effect of urbanisation on flood flows in Oak Creek, WI.

The increased magnitude and frequency of high flows has major adverse effects on the community located near the watercourse, on the floodplain, and on the ecology of the urban stream.

Most urban watershed management projects in the United States are driven by flood control objectives. On one side, public media pay extraordinary attention to the plight of people affected by flooding, resulting in heavy pressure and lobbying of public officials by affected individuals and citizens groups. However, using traditional benefit/cost analysis, most urban flood control projects are highly inefficient. In the Milwaukee (Wisconsin) metropolitan area, the benefit/cost ratios of flood control drainage projects in which the benefit is the reduction of monetary damages to properties and land within the floodplain, are typically less than 0.2. Consequently, projects that address flood control only are not feasible. Such projects would represent a massive transfer of benefits from the general taxpayer public to a small number of beneficiaries located in the floodplain. Furthermore, the *antidegradation rule* of the present regulations in the United States and elsewhere does not allow downgrading the integrity of the receiving water bodies even when the objective is drainage or flood control. Therefore, the sometimes conflicting concurrent objectives of drainage/flood control and restoration of ecological integrity of urban streams projects must be reconciled.

Restoration of ecological integrity of urban streams, on the other hand, benefits much larger segments of the population. However, considering the benefits of ecological improvement in the classic economic benefit cost analysis is difficult because such benefits are mostly intangible. On the other hand, such benefits are desired by the public, especially those living near the water body but not in the floodplain. Consequently, another measure of benefits must be defined and substituted. The *willingness to pay* of the public for the ecological benefits is a common substitution for strictly monetary flood control benefits in a multi-objective watershed restoration and flood control projects. In reality, projects that include both flood control benefits and ecological and water quality restoration and improvement may become acceptable to the general public as exemplified in two such projects in the Milwaukee metropolitan area that are featured in this article.

3.   Risks of Floods and Ecological Integrity

In order to compare the risks of increased flooding and the deteriorated ecological integrity of urban streams, the first step is to define a measure for both. Ideally, these measures should be comparable but, at this point, this seems impossible. The next best solution is to assign weights so they can be compared.

3.1.   MEASURE OF FLOOD RISK

There is a need to express a flood risk relation in the urban floodplain. First, let us define a flood as a flow that is greater than the capacity flow of the channel. A floodplain is a part of the river corridor (Figure 2).

It is also necessary to expand the probabilistic definition of flooding to areas away from the channel. As one moves away from the river's edge (the beginning of the floodplain) the probability of flooding decreases and at some point at a distance X from the river's edge the recurrence interval of flooding becomes 100 years, i.e., the risk of flooding is $r(X) = 0.01$. This is the extent of the 100 year floodplain as defined and delineated for engineering and flood insurance purposes. The schematic of the risk is then shown on Figure 2. If before urbanisation the smallest flow that leaves the channel is approximately a flow with a recurrence interval of 2 years (Figure 2), then the annual risk of flooding at the bank of the river is $r_n(0) = \frac{1}{2} = 0.5$. If, as a result of urbanisation, flooding becomes more frequent, for example, if the bankfull capacity flow is exceeded twice a year, the risk of flooding at the river's bank becomes $r_u(0) = 1/0.5 = 2$ and so forth. The subscripts *n* and *u* denote natural (pre-development) and urbanised (post development) conditions, respectively.

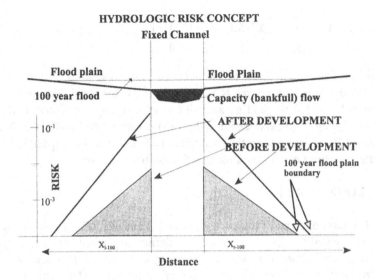

Figure 2. Concept of urban flood risk.

The monthly probability (risk) can be calculated from a *series* of maximum monthly flows and not just from one per-year maximum flow. This approach considers the fact that there may be more than one occasion in a year when the flow leaves the confines of the channel and becomes a flood. Thus, the monthly risk will be slightly different from 1/12 of the annual risk that is based on only one flood per year. To bring the magnitude of the risk on par with the water quality risks that are expressed in terms of the probability of daily grab or four day composite samples exceeding the acute or chronic toxic concentrations or water quality criteria, the risk of a monthly flood would be further divided by 30.42 (i.e. 365/12).

The logarithmic form of the risk function is selected for convenience and simply expresses the fact that floods on rare occasions may extend further than the 100-year floodplain limits. The risk function can then be expressed as

$$r = C\,10^{-K\,x} \tag{1}$$

The function parameters can be easily estimated from the knowledge of the risk of exceeding the bankfull capacity flow and from the extent of the 100-year flood plain. In the above equation, C is the risk of exceeding the bankfull flow, or, $C = r(0)$. In the Geographical Information Systems (GIS) environment, the risk function can be ascertained from flood flow elevations and contours of the floodplain. This risk function can be integrated, *i.e.*,

44

$$R = \int_0^\infty r_l(x)\,dx + \int_0^{+\infty} r_r(x)\,dx = r(0)\int_0^\infty [\,10^{-K_l\,x_l} + 10^{-K_r\,x_r}\,]\,dx \quad (2)$$

where subscripts *l* and *r* correspond to left and right bank floodplains.

The floodplain risk parameter, **R**, or function, **r**, can be combined with the flood damage cost information to yield an annualised flood damage indicator. If **d** is an uniform flood damage cost expressed in dollars ($) per m² of the flood plain, then the total annualised flood damage function is simply D = R x d. It will be argued in section 4.0 that better measures of flood control benefits can be developed.

3.2.   ECOLOGICAL RISK

Following US Environmental Protection Agency [2] and WERF [3] risk assessment documents, ecological risk for aquatic systems is defined as *"a probability that a genus residing in or potentially indigenous to the receiving water body will be lost or acutely damaged by existing or potential discharges of pollutants."* The term *potentially indigenous* reflects the fact that the representative composition of organisms should be selected from a composition in similar unimpacted water bodies located in the same ecoregions.

The calculations of individual risks for each stressor are demonstrated in Figure 3. Novotny and Witte [4] expanded the WERF[3] methodology to consider estimating the ecological effects of the wet weather (stormwater) and dry weather flows separately. EPA currently evaluates ecological risks in terms of the loss of species or genera that will result from the environmental impact [2,3]. This risk is basically a joint probability function of (1) probability density function of concentrations, **f(EMC)**, and (2) probability that species will be lethally or chronically impacted when exposed to a given concentration, **g(R | EMC)**. A simple model and method for calculating ecological risks of contaminants present in stormwater discharges was published by Novotny and Witte [4]. The method assumes that the event mean concentrations of pollutants are log-normally distributed. At this point, the method estimates only the risk of acute damage to the indigenous population. Both stormwater and base flow discharges are considered. The method considers dilution of stormwater and CSO discharges and the water effect ratio. A simple software package has been developed [5]. The single, dimensionless risk value has numerous advantages over the traditional separate comparison of measured water quality data with criteria because it puts all pollutants on the same basis, *i.e.*, the probability of ecological damage to the resident biota (or potentially resident as derived from reference unimpacted water bodies of the same character within the ecoregion). It may also be an additive and comparative number, *i.e.*, risks from several compounds and those from dry weather discharges could be added together to yield an overall risk and approximate synergy and individual risks can be quantitatively compared.

The overall ecological state of the receiving water body can be ascertained using biological evaluation, such as that outlined in the *Rapid Bioassessment Protocol* methodology [6]. However, biological assessment procedures of this type are based on the application of multiple indices calling for subjective judgement and mix together symptomatic and causative parameters. For example, a biotic index based on fish or microinvertebrate composition is symptomatic while chemical concentrations and physical habitat parameters are causative. An overall relationship between causative risk and symptomatic IBI parameters and root cause of impairment are shown on Figure 4.

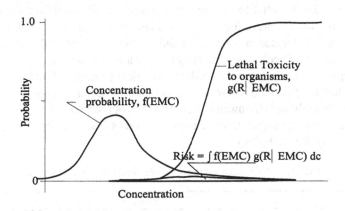

Figure 3. Concept of risk calculation for an individual stressor.

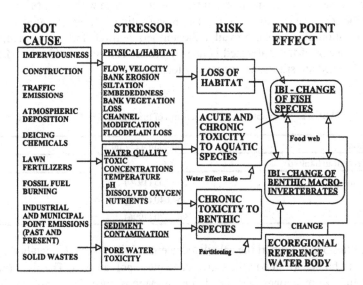

Figure 4. Causative and end point symptomatic components of the water body integrity.

4. Investigated watersheds

The methodology presented in this paper has been applied to two watersheds, the mostly rural but rapidly urbanising Oak Creek watershed and the developed Menomonee River watershed.

The Oak Creek watershed (see Table 1) discharges into Lake Michigan in the City of South Milwaukee, WI. The Oak Creek watershed can be characterised as mostly rural with a great potential for future development. Agricultural land (cropland and pasture) represents the prevailing type of the land use in the watershed. Most of the agricultural land is located in the western and southern portions of the watershed. The soils within the Oak Creek watershed are silty clay loams, loams, and sandy loams, and are developed on gently sloping or rolling moraine topography. Most of the soils are relatively fertile. Pollution sources can be categorised as municipal, industrial, agricultural, landfill, and stormwater. A contribution of pollution from the point sources is negligible compared to that from the nonpoint sources. Rural sources dominate among the nonpoint sources (20-50%).

TABLE 1. Basic watershed characteristics of the Oak Creek watershed.

Area	69.8 km^2 (27.2 mi^2)
Percent urbanised	44.6%
Population (1980)	39700

The Menomonee River (see Table 2) watershed discharges into the Milwaukee River about one and half km upstream from its confluence with Lake Michigan. About 45% of the total area is still in rural use, representing a great potential for nonpoint source pollution. Rural areas prevail in the northern portion of the watershed, while the southern portion of the watershed is mainly urban. The soils within the Menomonee River watershed are silt loams or gravelly loams.

TABLE 2. Basic watershed characteristics of the Menomonee River watershed.

Area	350.7 km^2 (137 mi^2)
Percent urbanised	52.8%
Population	348,165 (1970)
	964,640 (1990)
	962,570 (1996)

Tables 3 and 4 report the calculated chemical risks by toxic metals for two pilot urban water bodies located in the Milwaukee metropolitan area. The risks (both acute and chronic) from copper calculated for the Menomonee River are two orders of magnitude higher than those for the Oak Creek. The risks from lead and zinc are at the same level for both watersheds. A risk greater than 10^{-4} may indicate impairment of biotic integrity and a loss of genera.

The acute risks in the Menomonee River are generally higher, reflecting a higher degree of urbanisation. Chronic risks are about the same in both water bodies. Because both acute and chronic risks represent the same impact, the loss of genera, one may conclude that the chronic risks affect the integrity of the water body more than the acute risks.

TABLE 3. Chemical acute risk to aquatic biota. Oak Creek and Menomonee River.

River and station identification	Cu	Pb<1987	Pb>1987	Zn
Oak Creek - RI-23	0	0.00021	0.00001	8.8.E-04
- RI-24	0	0.00018	0	0.00098
- RI-25	0	0.00015	0	0.00087
- RI-26	0	0.00014	0	0.00088
- RI-27	0	0.0002	0	0.00084
Menomonee River - RI-16	0.00002	0.00028	0	0.00066
- RI-21	0.00003	0.00031	0	0.0011
- RI-22	0.00003	0.00024	0	0.0013
- RI-09	230000	0.00018	0	0.0014
- RI-10	0.00003	0.00015	NA	0.0018

TABLE 4. Chemical chronic risk to aquatic biota. Oak Creek and Menomonee River.

River and station identification	Cu	Pb<1987	Pb>1987	Zn
Oak Creek - RI-23	0	0.028	0.0034	0.0017
- RI-24	0	0.028	0.0025	0.0019
- RI-25	0	0.026	0.0026	0.0014
- RI-26	0.00001	0.026	0.0021	0.0017
- RI-27	0	0.028	0.0021	0.0016
Menomonee River - RI-16	0.00016	0.034	0.0021	0.0021
- RI-21	0.00023	0.04	0.002	0.0021
- RI-22	0.00024	0.036	0.0076	0.0025
- RI-09	0.00002	0.029	0.00076	0.0026
- RI-10	0.00022	0.03	NA	0.0035

5. Willingness to Pay for Reducing Flood Risk and Ecological Damage

To assess community support for reductions in ecological risks or flood control, a direct valuation approach known as Contingent Valuation (CV) was employed. The CV method employs surveys of residents within the watershed to gauge individuals' willingness-to-pay for nonmarket goods. The willingness-to-pay (WTP) concept reflects the maximum amount of money that an individual is willing to sacrifice for an improvement in a public good, compared to the status quo. Responses can be interpreted as the dollar payment that would make the person indifferent between the two states of the world and hence the stated WTP represents a true economic measure of the maximum individual benefit that can be derived from the public project. Since WTP places a monetary value on a good, it can be used to determine the relative weight of project attributes. Specifically, it can be used to determine how flood control objectives compare to environmental objectives of watershed management projects. Once the relative values of hypothetical projects can be derived, then community-wide benefit estimates can be compared with project costs to determine optimal policy [6-11]. The methodology has also been applied to valuations of flood control [12-14]. If carefully designed, these studies can capture a range of different types of benefits. For example, flood control will probably be highly valued by those living in areas that experience frequent flooding, and it may also be of value to residents of the watershed who do not personally experience flooding, but who favour community flood control for other reasons such as altruism. Likewise, some residents may have WTP for ecological restoration projects because they use the river for recreational purposes (i.e., use values), whereas others may believe that environmental restoration is beneficial in its own right (i.e., nonuse values).

5.1. RELEVANT DRIVERS OF WTP

To derive benefits from improvement in either ecological risk or flood risk, WTP must be correlated to the relevant drivers. Traditionally, economists focussed exclusively on demographic and economic factors. These include factors such as respondent income, race/ethnicity, gender, owner/renter status, distance from the resource (e.g., floodplain or river), and family size. More recently, however, economists have begun to team with social scientists from other disciplines to understand the role that psychological, social, and communication variables play in determining WTP. In this case, we start by applying one of the most successfully-tested models in psychology, Ajzen's [15] Theory of Planned Behaviour (TPB), to the task of understanding the likely reasons that individuals would be willing to pay various amounts (including nothing) for watershed improvement and flood control projects. In particular, the model is used as a framework for examining how individuals' beliefs about the costs and benefits of supporting a watershed project combine with how they value those cost-benefit outcomes to potentially affect their willingness to pay for the project. Such an examination can provide the basis for understanding some of the dynamics of public opinion about publicly-supported projects such as these as well as form the basis for planning public information and education efforts.

The Theory of Planned Behaviour is applied in this study as a means of determining some key predictors of WTP. In this study, WTP is considered to be a form of *Behavioural Intention* found in past studies to be an important predictor of actual behaviour. Ajzen's theory indicates that behavioural intention (BI) is predicted by a limited set of psychological variables whose predictive power relative to each other can vary from behaviour to behaviour:

I. *Perceived Behavioural Control* (PBC), which is one's sense of control over the behaviour (*e.g.*, the amount one could easily pay and/or the extent one has control over the amount paid);

II. *Subjective Norms* (SN), which are one's social normative beliefs (*e.g.*, one's sense that other people important to the individual would want him or her to pay for the benefit);

III. *Cognitive Structure* (also known as *Indirect Attitude*) which is composed of a set of beliefs about the cost-benefit consequences of performing the behaviour (*e.g.*, that paying a given amount would in fact help people who live in the floodplain), and a set of values that the individual holds about those consequences (*e.g.*, that helping people who live in the floodplain in this way is a desirable outcome). The latter two elements (outcome beliefs and outcome evaluations) are considered to be the building blocks of an *attitude toward performing the act* (AAct), which is the more direct predictor of BI.

5.2. WTP ESTIMATE OF STABILITY

The stability versus volatility of WTP estimates can be of concern to watershed program planners who rely on such estimates, usually gathered from surveys, as guides to what the public will support in terms of the costs of such projects. Unreliable and unstable WTP estimates can be the result of methodological flaws or of real changes among a significant portion of the public and can deceive planners who might find that public support as tapped through a WTP survey at one point in time has, to their surprise, eroded by the time financing for the project comes to fruition later. Thus, this project uses TPB, coupled with insights derived from the Heuristic-Systematic Model of human information processing [16,17], to examine some precursors to the stability or volatility of WTP estimates.

Following from Ajzen and Sexton [18], cognitive structure, AAct, and eventually willingness to pay will probably be more stable over time when persons process issue-relevant information systematically (i.e., more deeply and with full effort) rather than heuristically (i.e., superficially). For example, individuals who process risk-related information more systematically have been found to take more behavioural beliefs into account when deciding how to behave personally in the face of an environmental health risk [19]. It is likely that their resulting cognition, attitudes, and intention to behave in a given way in the face of a risk are relatively resistant to change as compared to those persons who process risk information more superficially. Our intent is to study the stability of these relationships by comparing them over both waves of the panel design survey.

50

5.3. METHODOLOGICAL ISSUES

A 25-minute telephone survey was conducted on nearly one thousand randomly selected adult residents of two metropolitan Milwaukee watersheds, Menomonee River and Oak Creek, in the late fall and winter of 1999-2000. The survey was conducted by a professional survey research organisation, the University of Wisconsin Survey Center at the University of Wisconsin-Madison, in two waves.

The survey organisation also conducted eight carefully constructed focus groups in spring 1999, prior to the survey, to help in the development of the survey instrument. The focus groups were organised along three different dimensions: watershed of respondent (Menomonee River vs. Oak Creek), upstream vs. downstream location, and environmental vs. flood questions. Each focus group included 5-8 randomly-chosen individuals and a facilitator. The goal of the focus groups was not to generate conclusions on WTP, but rather to assist in the development of survey items, which included the testing of the wording of key items as well as the conceptual development of key variables.

In both waves, respondents are asked the same set of questions and are provided with a detailed description of the watershed project before being asked a series of questions designed to determine their willingness to pay. The WTP questions were posed in terms of a hypothetical political referendum. Respondents were asked to indicate the maximum amount of money (if anything) that the plan could require them to pay annually for the next 20 years yet still allow them to vote in favour of the plan.

The objectives of the survey areas follows.
1. Distinguish between the WTP for flood-control and the WTP for ecological restoration of urbanising watersheds and determine whether the two types of benefits are separable.
2. Test a model that describes the salient drivers (psychological, sociodemographic, locational, and communication-related) that appear to influence WTP responses in the survey.
3. Determine whether WTP responses are stable over time.
4. Relate WTP to the underlying flood and ecological risk improvements in the project and then derive estimates of community-wide benefits from flood control and/or ecological risk reduction in the watershed. Benefits are then compared with project costs to determine optimal watershed policy.

This paper includes a summary of the relevant issues that must be confronted when designing a CV survey to be used for watershed management purposes, and then give a peek at some of the preliminary findings as they relate to the first two objectives listed above.

5.4. SURVEY DESIGN

5.4.1 *Definition of the Good*
The first task for the socio-economic team was to work with the engineering team to determine how flooding is likely to worsen in the absence of additional public spending, and also to describe the improvements in ecological health that were possible. It was

determined that, without additional public spending, the recurrence level of flooding would increase substantially in the Menomonee River and Oak Creek due to increased flows and expansion of the floodplain. Furthermore, biological studies of both watersheds indicated that chronic exposures would diminish water clarity, reduce the variety of fish and wildlife species, and increase the levels of toxic chemicals in the stream or river.

The second task was to determine the appropriate manner in which to describe these potential changes in risks resulting from urbanisation. The focus groups were conducted in part to accomplish this second task. The focus groups revealed some interesting insights.

- Although hydrologic simulations showed that continued urbanisation in the Oak Creek area would likely generate substantial increases in flood risks and an expanding floodplain, few Oak Creek participants in the Oak Creek focus group sessions believed the threat was likely to occur. Thus, it was determined that Oak Creek residents would be asked questions only about ecological quality.
- Focus group participants had difficulty distinguishing between probability of flooding and meaningful recurrence intervals in light of three "100-year floods" over the period 1986 - 1998. It was determined that the best way to phrase the question was to describe (1) the relative and actual increase in flood risk (i.e., 3-5 times higher risk, and the actual likelihood of flooding increased from 3%-5%) rather than describe the recurrence interval. In addition, to capture how the floodplain was likely to expand with urbanisation, we determined the likely number of additional homes that would be flooded as a result of urbanisation, assuming no action were taken by watershed management officials.
- To convey likely improvements in the ecological health of the system, it was determined that it was best to describe water clarity, presence of toxins in the water (without going into detail about how the toxins got there in the first place) which would make wading and swimming in the stream or river potentially unhealthy, and the variety of wildlife and fish around the river.

5.4.2. *The Problem of Embedding*

One of the potential problems with CV studies results when the magnitude of the WTP response is independent of the level or scope of improvements being considered [20]. In the context of this study, the embedding issue is related to whether WTP for the good increases when both flood control and ecological improvements are being described, versus the situation where only flood control or only ecological quality is being valued.

To investigate this possibility, three different WTP questions were developed. The first type of question asks one set of respondents their WTP for flood risk only; the second asks another set of respondents their WTP for ecological improvement, and the final question asks a third set of respondents their WTP for both flood risk and ecological improvement. If embedding is a problem, we would expect that WTP would be higher for the combined flood risk and ecological improvement than for questions with one or the other goods, but not both.

5.5. RESULTS

In all, 999 respondents were interviewed in the first wave, 303 (all from the Menomonee River watershed) were interviewed about their willingness to pay for a flood control project on that river (which we dubbed the "flood path" of questions), 459 (from both the Menomonee River and Oak Creek watersheds) were interviewed about their WTP for an ecological restoration project in their respective watersheds (the "environmental path" of questions), and 237 (all from the Menomonee River watershed) were interviewed about their WTP for a combined project that would hold the line on flooding as well as improve ecological quality of the watershed (the "combined path" of questions). Due to interviewing time limitations, only those respondents in the flood and environmental paths of questions were asked a full battery of questions based on the psychological and communication models.

Initial results indicate that there are no statistically significant differences between paths in the maximum amounts individuals would be willing to pay for the projects (overall mean=$84), even though the combined project would deliver more benefits than either of the two individual projects (flood control and ecological improvement) (Figure 5). The non-significance remains when the results are adjusted for respondent income, education, race/ethnicity, gender, age, distance from the river/creek, and the number of inhabitants of the dwelling. This suggests the possibility of the embedding problem identified by Kahneman and Knetsch [20] is present in these findings. Once the second wave of the survey is completed, a more complete assessment of this potential problem will be conducted.

In addition, WTP for the flood control project and for the ecological restoration project is more strongly associated with the psychological predictors from the Theory of Planned Behaviour, especially with cognitive structure, than with income and the other sociodemographic predictors. As an example, we turn our attention to the individual variables associated with WTP for the flood control project, since that project includes an element of potential environmental enhancement which might contribute to WTP for the project.

The primary sociodemographic variables (respondent income, education, race/ethnicity, gender, age, dwelling location within the floodplain, and the number of inhabitants of the dwelling) bear weaker relationships with WTP for the flood control project than do a set of variables based on the Theory of Planned Behaviour, specifically, subjective norms ($r=.29$, $p<.001$) and an overall index of cognitive structure ($r=.40$, $p<.001$; $r=.46$, $p<.001$, when the belief-evaluation compound items are also multiplied by a separate self-report measure of the importance of the outcome to the decision). Among the belief-evaluation compounds themselves, correlations with willingness to pay for the flood control project are as follows (the second coefficient representing the addition of the importance multiplier):

- Add significantly to my taxes ($r=.03$, ns; $r=.11$, $p<.05$);
- Be personally expensive for me ($r=.15$, $p<.01$; $r=.21$, $p<.001$);
- Make me feel like I am doing something for the environment ($r=.28$, $p<.001$; $r=.34$, $p<.001$);

- Make me feel like I am doing something for the community (r=.34, p<.001; r=.45, p<.001);
- Probably help support a long term solution (r=.36, p<.001; r=.39, p<.001);
- Probably help future generations (r=.32, p<.001; r=.36, p<.001);
- Probably help to hold the line against flooding (r=.28, p<.001; r=.33, p<.001) ;
- Probably help people who live in the flood plain (r=.19, p<.001; r=.28, p<.001)

A microscopic analysis of the responses indicates that about 73% of the respondents believe that the project would produce a *bad* outcome of helping people who live in the floodplain. Only about 4% see the project as helping people who live in the floodplain and put a positive value on that outcome, which suggests that altruistic motives are certainly not strongly at work in the watershed population.

A microscopic analysis of responses concerning whether the project would make the individual feel like he/she is doing something for the environment, one of the stronger variables in terms of correlation with WTP, illustrates that about 73% of the respondents believe that paying the amount they specified for the flood control project would provide this sense of environmental support as a good (positively valenced) outcome for themselves. However, almost 16% of the respondents also put this kind of positive value on feeling that they are doing something for the environment yet either believe that the project would not deliver that outcome or are unsure as to whether it would or not. Presumably, if they were to change their beliefs about the project along this dimension, their support for it might increase to some extent. Only a small portion of respondents (3%) feel negatively toward feeling that they are doing something for the environment and also believe that the project is unlikely to deliver that kind of support. An almost identical pattern of responses exists for beliefs and evaluations about having a sense of doing something for the community. Thus, if public education efforts were to help convince more members of the public that support for the project produces a sense of helping out the environment or the community, or raises the salience (importance) of those beliefs as decision factors in supporting a flood control project, the model would predict a net gain (albeit not huge) in support via WTP for the project. As noted previously, however, it should be assumed that such beliefs are necessarily easy to affect.

6. Conclusions and Future Directions

It is clear that traditional methods for assessing economic benefits from watershed management policy alternatives are inadequate. By focussing exclusively on the direct benefits from flood control (i.e., damage prevention) while ignoring more indirect benefits related to altruism and environmental improvement, traditional approaches send false signals as to the societal benefits of different policy alternatives. The interdisciplinary approach outlined above draws upon recent advances in the natural sciences to accurately assess the risks associated with alternative flood control projects. Accurate descriptions of risks are essential for efficient policy decisions. Technical risk estimates are then conveyed in lay terms that watershed residents can comprehend, and

54

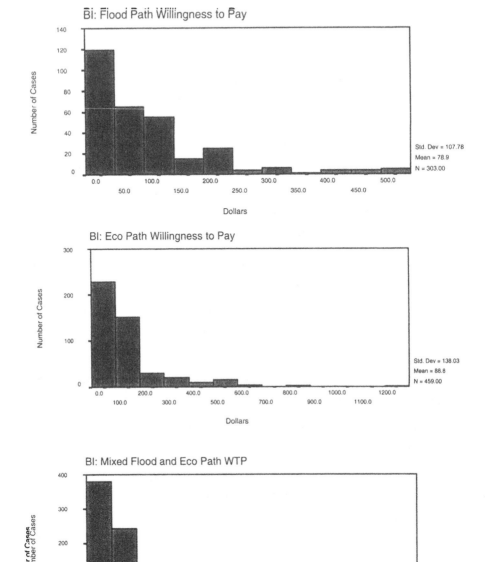

Figure 5. Survey results.

then the CV approach is used to determine the separate WTP for flood control and ecological restoration within urban watersheds. Preliminary empirical work reveals that both economic and psychological drivers appear to be important determinants of WTP for flood control and the ecological restoration of urban watersheds.

The next phase in this research will be to determine the relationship between the WTP response and the individual flood risk and ecological risk exposure so that total watershed benefits can be derived. In addition, a second wave of the survey will permit a more in-depth investigation of the stability of WTP responses. It is our belief that the use of this approach will lead to more informed policy debates and better decisions.

7. Acknowledgment and Disclaimer

The research described in this paper was sponsored by the US Environmental Protection Agency's STAR watershed program. The sponsorship is greatly appreciated. The findings and conclusions of this paper are those of the authors and not of the funding agency.

8. References

1.   McCuen, R.H. (1998) *Hydrologic Analysis and Design,* Prentice Hall, Upper Saddle River, NJ.
2.   US EPA (1992) Framework for ecological risk assessment, Risk Assessment Forum, EPA 630/R-92/001, U.S. Environmental Protection Agency, Washington, DC.
3.   Parkhurst, B.R. Warren-Hicks, W., Cardwell, R.D., Volosin, J., Etchinson, T., Butcher, J.B., and Covington, S.M. (1996) Aquatic ecological risks assessment: a multi-tiered approach. Report 91-AER-1, Water Environment Research Foundation, Alexandria, VA.
4.   Novotny, V. and Witte, J.W. (1997) Ascertaining aquatic ecological risks of urban stormwater discharges, *Water Research* **31(10)**:2573-2585.
5.   Bartošová, A. (2000) Statistical considerations in aquatic ecological risk estimation, Technical Memorandum # 1. Institute for Urban Environmental Risk Management, Marquette University, Milwaukee, WI.
6.   Barbour, M..T., Gerritsen, J., Snyder, B.D., and Stribling, J.B. (1997) *Revisions to* rapid bioassessment protocol for use in streams and rivers: peryphyton, benthic macroinvertebrates, and fish. EPA-841/D-97/002, U.S. Environmental Protection Agency, Washington, DC.
7.   Randall, A., Ives, B.C., and Eastman, C. (1974) Bidding games for valuation of aesthetic environmental improvements, *Journal of Environmental Economics and Management* **1**, 132-149.
8.   Eastman, C., Randall, A., and Hoffer, P.(1978) A socioeconomic analysis of environmental concern: Case of the four corners electric power complex, Bulletin 62, Agricultural Experiment Station, University of New Mexico, Albuquerque.
9.   Acton, J.P. (1973) Evaluating public programs to save lives: the case of heart attacks, Research Report R-73-02, Santa Monica, Rand Corporation.
10.   Hanneman, M. (1978) A methodological and empirical study of the recreation benefits from water quality improvement, Ph.D. dissertation, Harvard University.
11.   Binkley, C.S. and Hanneman, W.M. (1978) The recreation benefits of water quality improvement: Analysis of day trips in an urban setting, Report to the U.S. Environmental Protection Agency Washington, DC.
12.   Thunberg, E.M. (1988) Willingness to pay for property and nonproperty flood hazard reduction benefits: An experiment using the contingent valuation survey method, Ph.D. dissertation, Virginia Polytechnic Institute and State University; Ann Arbor, UMI Dissertation Services.
13.   Shabman, L. and Stephenson, K. (1996) Searching for the correct benefit estimate: evidence for an alternative perspective, *Land Economics* **72** (Nov), 433-449.

56

14. Johnson, R.J., Swallow, S.K., and Weaver, T.F. (1999) Estimating willingness to pay and resource tradeoffs with different payment mechanisms: an evaluation of a funding guarantee for watershed management, *Journal of Environmental Economics and Management* **38**, 97-120.

15. Ajzen, I. (1988) *Attitudes, Personality, and Behavior*, Open University Press, Milton Keynes, UK.

16. Eagly, A.H. and Chaiken, S. (1993) *The Psychology of Attitudes,* Harcourt Brace Jovanovich, Fort Worth.

17. Griffin, R.J., Dunwoody, S., and Neuwirth, K. (1999) Proposed model of the relationship of information seeking and processing to the development of preventive behaviors, *Environmental Research* **80:** S230-245.

18. Ajzen, I. and Sexton, J. (In press) Depth of processing, belief congruence, and attitude behavior correspondence, in S. Chaiken and Y. Trope, Eds. *Dual Process Theories in Social Psychology.* Guilford, New York.

19. Griffin, R.J., Neuwirth, K., Giese, J., and Dunwoody, S. (1999) The Relationship of Risk Information Processing to Consideration of Behavioral Beliefs, Presented to the Science Communication Interest Group, Association for Education in Journalism and Mass Communication, annual convention. New Orleans LA. August.

20. Kahneman, D. and Knetsch, J. (1992) Valuing public goods: the purchase of moral satisfaction, *Journal of Environmental Economics and Management* **22:** 57-70.

STRATEGIES FOR MANAGEMENT OF POLLUTED STORMWATER FROM AN URBAN HIGHWAY IN GÖTEBORG, SWEDEN

G. SVENSSON*, P-A. MALMQVIST**, & S. AHLMAN*
*Water Environment Transport, Chalmers University of Technology,
S-412 96 Göteborg, Sweden
**Urban Water, Chalmers University of Technology,
S-412 96 Göteborg, Sweden

1. Introduction

Stormwater from urban highways contains high concentrations of heavy metals, polycyclic aromatic hydrocarbons (PAHs) and other substances that are harmful to the environment. The choice of abatement strategies is not trivial. If the stormwater is brought to the municipal wastewater treatment plant (WWTP) in combined sewers, it will cause overflows at combined sewer overflow (CSO) points, accumulation of contaminated sludge that cannot be used as fertilisers on farmland, and sometimes disturbances of the operation of the WWTP. If the stormwater is discharged to a receiving water it will carry all the pollutants directly to the water body, most often a small urban creek not suitable for such discharges. If the stormwater is allowed to infiltrate, the groundwater and in some cases the vegetation will be effected. In an action plan for the improvement of the stormwater handling all these issues have to be taken into account. Abatement of pollutants at the source is one very important strategy that should have first priority. Thus, to improve the receiving water in a sustainable way the sources of the pollution have to be identified, quantified and reduced.

A conceptual model for describing and quantifying the stormwater pollutant sources and for simulation of the effects of abatement measures has been developed and used for the planned area Hammarby Sjöstad in Stockholm [1]. It has also been verified against actual measurements for the Lake Trekanten catchment area [2].

A mathematical model for describing and quantifying the stormwater pollutant sources and for simulation of the effects of abatement measures has been developed and used for the "Järnbrott Catchment" [3] and [4]. The model was considered useful for understanding the processes of generation and transport of the studied substances. The "Järnbrott Catchment" has been monitored extensively, as reported elsewhere [5]. The model has been verified against measured data. Six substances are included in this

57

J. Marsalek et al. (eds.), Advances in Urban Stormwater and Agricultural Runoff Source Controls, 57–67.
© 2001 Kluwer Academic Publishers. Printed in the Netherlands.

58

study: lead (Pb), cadmium (Cd), copper (Cu), zinc (Zn), phosphorus (P) and nitrogen (N).

2.   Catchment area "Järnbrott", Göteborg

The catchment area "Järnbrott" consists mainly of an urban highway and is located in the southern parts of the City of Göteborg. The urban runoff from the catchment is discharged to a wet pond that was constructed in 1996. The impervious catchment area was estimated at 160 ha by the municipality, but the observed runoff from the catchment corresponds to 220 ha (see TABLE 1). The pond has a water surface area of 6200 m^2 in dry weather. The depth of the pond varies between 0.5 m and 1.6 m in dry weather. All stormwater from the area drains to the pond until the inflow exceeds about 700 l/s. When this occurs, the overflow starts diverting the excess part of the stormwater directly to the Stora Ån river. Maximum inflow to the pond is estimated as about 1100 l/s when the total inflow to the overflow reaches the maximum of about 8 m^3/s. Due to the overflow, about 80% of the annual stormwater load is treated in the pond before reaching the river.

TABLE 1. Input data for the catchment area.

Total impervious area	2,200,000	m^2
Roads	1,161,500	m^2
Zinc surfaces by roads	5	%
Roofs	423,600	m^2
Zinc roofs	25	%
Copper roofs	20	%
Other impervious area	614,900	m^2
Total vehicle km	140,000	km/day
Heavy vehicles	4	%

The catchment runoff has been observed using automatic flow-proportional stormwater samplers at the inlet and outlet. At the outlet the discharge was monitored by a pressure probe, measuring the pond level above the outlet weir. A raingauge was installed at the site. All field data were stored in the samplers by a built-in data logger from which data were "downloaded" after each storm event by a laptop computer.

Laboratory analysis of the stormwater samples included: suspended solids, total (TSS) and volatile (VSS), heavy metals (zinc, copper, lead and cadmium) and nutrients (total nitrogen and phosphate-phosphorus). Suspended solids concentrations in stormwater were determined according to a Swedish standard method (SS 02 81 12). TSS were analysed by filtering the stormwater through a GF/C glass fibre filter. Heavy metals (zinc, copper, lead and cadmium) were determined by an atomic absorption spectrophotometer (SS 02 81 52-2 and SS 02 81 84-1). Nutrients (total nitrogen and phosphate phosphorus) were analysed by spectrophotometric methods, SS 02 81 31-1 and SS 02 81 26-2, respectively.

The stormwater quality measurements included two periods, from August 1997 until February 1998, and from April 1998 until July 1998. A summary of EMCs for the observed constituents are shown in TABLE 2.

TABLE 2. Range of pollutant EMCs and long term removal efficiencies for the pond and for the whole system considering the overflows.

Pollutant	Inflow		Outflow		Removal efficiency (%)	
	Range	Mean	Range	Mean	Pond	System
TSS (mg/l)	6.4- 820	55	6.0 – 33	17	70	41
VSS (mg/l)	1.6 - 180	16	1.0 – 11	6.5	60	39
Zinc (μg/l)	42 - 520	120	9.0 – 180	81	31	24
Copper (μg/l)	16 - 210	53	18 – 72	37	30	23
Lead (μg/l)	2.1 - 77	13	1.4 - 16	6.8	48	29
Cadmium (μg/l)	0.16 - 1.3	0.55	0.20 - 1.0	0.48	11	11
Nitrogen (mg/l)	0.63 - 5.3	2.0	0.82 - 3.6	1.9	7	8
PO_4-P (μg/l)	20 - 560	70	16 - 89	42	40	27

3. Model layout and description

A conceptual model, *Figure 1*, was developed, based on models used in earlier studies in Stockholm [1,2].

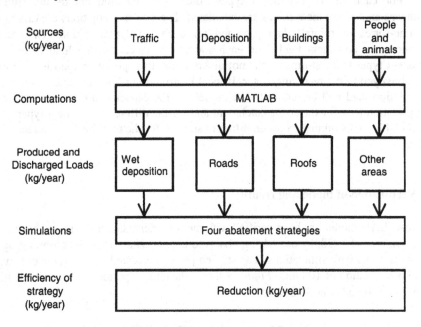

Figure 1: Conceptual source model

The simulations were carried out with a model in the MATLAB/Simulink called SEWSYS. The model is derived from a diploma work at Chalmers University of Technology [4]. SEWSYS is a model for materials flow in sewer systems, in which both separate and combined systems can be simulated. The time step is variable, here 5 and 15 minutes were used.

For this study the stormwater module in SEWSYS was used, as shown in *Figure 2*. It includes a stormwater pollutant generator, runoff module and an optional pond for treatment. The pollutant generator was originally derived from a diploma work at the Uppsala University School of Engineering [3].

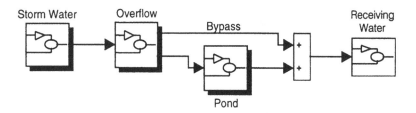

Figure 2. SEWSYS – Stormwater module

For each of the catchment types, total produced amounts of the studied substances and yearly mass fluxes are calculated. The model simulations are driven by the time-series of rainfall data provided for each simulation period. The model generates the constituent load from each source and calculates the load from each catchment type. If the stormwater is not diverted, the total pollution load is supposed to be conveyed to the receiving water. If local infiltration or local treatment is applied the pollution load will be reduced. The model has the capacity to display EMCs and total pollution load for the entire catchment and for the different catchment types. It is also possible to distinguish between different sources for the total load to the receiving water.

4. Verification of modelling results

The MATLAB-model has been verified with measurements taken in the Järnbrott area in the summer of 1998, by comparing observed copper, zinc, lead, cadmium, nitrogen and phosphorus with simulations. The studied period consisted of 17 rain events with a total precipitation of 108 mm. *Figure 3* shows the chosen period with rain intensities in μm/s distributed over 5 minutes.

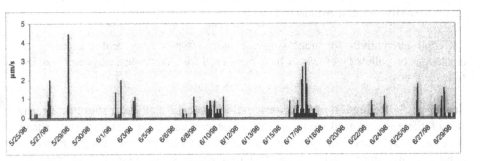

Figure 3. Järnbrott rain series (Summer 1998)

The results of the verification simulations showed an acceptable agreement with the measured pollution load (*Figure 4*). The largest differences were found for cadmium and phosphorus with errors of 81% and 73%, respectively. These high differences were presumed to be caused by the lack of data on the pollutant sources of these substances.

A reason for the tendency that the measured values were higher than the simulated ones could be due to the fact that the Järnbrott catchment area has a large amount of industrial land. Spills, etc. on industrial areas are not accounted for in the model.

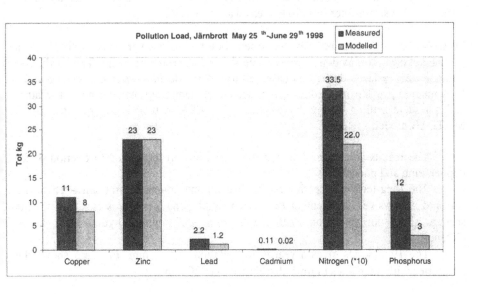

Figure 4. Verification results

5. Alternative abatement alternatives

Different alternatives for improving the stormwater quality and decreasing the discharges of polluted stormwater have been used for simulations, and combined into four scenarios:

Scenario 1 – Reduction of the construction material pollutant sources, e.g., the use of tile roofs instead of copper-plated or galvanised and painted steel sheets, painting of galvanised objects like lampposts and railings, and using concrete or natural paving stones instead of asphalt pavements. Simulations of these measures indicated an 80% decrease in copper and zinc surfaces, together with a 50% decrease in substances from asphalt.

Scenario 2 - Reduction of the mobile pollutant sources, e.g. better tyres, change of materials in brake linings, change of materials in the catalysts, less oil spills. This was simulated with a 50% decrease in substances from tyres, zero copper from brake linings, zero discharge from catalysts, 50% decrease in oil spills, along with a 30% decrease in traffic load.

Scenario 3 – Treatment by infiltration, involves measures taken within the stormwater system, i.e., infiltration of stormwater in open ditches. In simulations, it was assumed that all stormwater from roofs and 50% of other areas could be infiltrated. No stormwater from road areas was infiltrated.

Scenario 4 – Treatment in an open pond. Stormwater was drained to a sedimentation basin, as done, for example in the Järnbrott catchment. The particulate pollutants are reduced during the passage through the detention pond according to the EPA method [6]. The method takes into account dynamic settlings during wet weather and quiescent settling during dry weather. The EPA method is incorporated in the MATLAB model.

The results from the simulated four scenarios are presented in Section 6, for copper, zinc and phosphorus.

The long-term perspective is decisive for the choice of strategies. Therefore series of successive rain events were used in simulations. The rain series used for the scenario simulations was from 1926 and described well a normal hydrological year in Göteborg.

The 1926 series gives a yearly precipitation of 685 mm. *Figure 5* shows the rain series with intensities in μm/s distributed over 30 minutes.

63

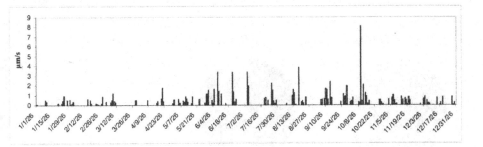

Figure 5. Rain series (1926) used for the scenario simulations

6. Results and discussion

By applying the MATLAB model to the conceptual model in *Figure 1*, the yearly mass flows for the six studied substances to the receiving water from a system with no measures have been calculated (TABLE 3). The table also shows the results of the scenario simulations for the four chosen scenarios. The results are explained in greater detail for copper, zinc and phosphorus.

TABLE 3: Pollution load to receiving water [kg/year]

	No measures	#1 Construction materials	#2 Mobile sources	#3 Infiltration	#4 Pond
Copper	83	41	57	30	57
Zinc	240	96	210	130	160
Lead	11.3	10.6	10.9	7.5	7.6
Cadmium	0.20	0.19	0.19	0.13	0.14
Nitrogen	1 800	1 800	1 800	1 200	1 200
Phosphorus	30	30	27	23	20

6.1 POLLUTANT SOURCES

In *Figure 6* the main pollutant sources for the six studied substances in the present system are shown.

The dominant sources of copper are roofs (roof plates and fittings) and roads (brake linings). The dominant sources of zinc are roads (tyres) and roofs (roof plates, fittings, lamp posts etc.). The dominant sources of phosphorus are wet deposition and roads.

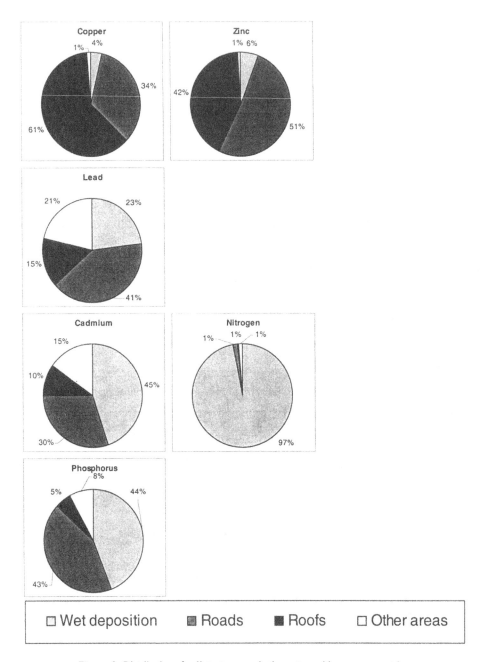

Figure 6: Distribution of pollutant sources in the system with no measures taken

6.2 RESULTS OF THE SCENARIO SIMULATIONS

The four scenarios described in Chapter 5 were simulated in a similar way as the present system. In TABLE 3 the results for six substances are shown. The results for the three selected substances are shown in *Figures 7-9.*

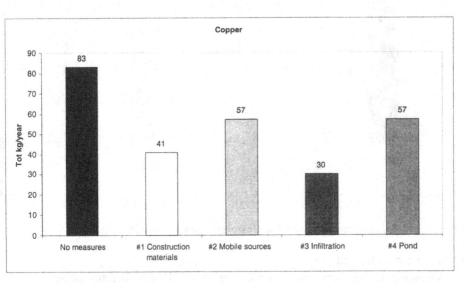

Figure 7: Copper mass flows to the receiving water in the four scenarios

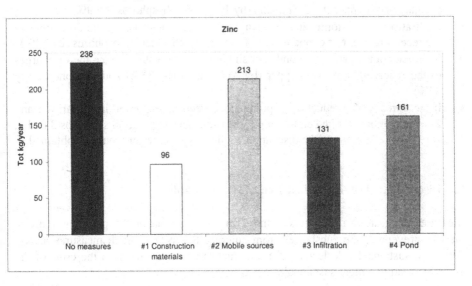

Figure 8: Zinc mass flows to the receiving water in the four scenarios

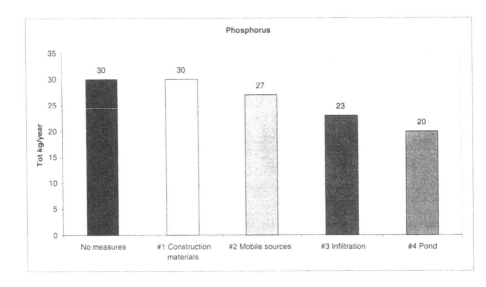

Figure 9: Phosphorus mass flows to the receiving water in the four scenarios

It may be concluded that

- Reducing the construction material sources would considerably decrease the discharges of copper (by 50%) and zinc (by 60%)
- Reducing the mobile sources (emissions from vehicles) would decrease the discharges of copper (by 32%), zinc (by 12%) and phosphorus (by 9%)
- Infiltration of the stormwater as a single measure would reduce the discharges to the receiving water of copper (by 64%), zinc (by 46%) and phosphorus (by 23%)
- Sedimentation in an open pond as a single measure would reduce the discharges to the receiving water of copper (by 31%), zinc (by 33%) and phosphorus (by 33%)
- If measures for the reduction of pollutant sources, as suggested in scenarios 1 and 2, were combined with "end-of-pipe" measures as suggested in scenarios 3 and 4, very high reductions of the discharges to the receiving water would be obtained.

7. Discussion: Are the studied scenarios sustainable?

Measures suggested in scenarios 1 and 2 imply reduction of the pollutant sources, thereby preventing the substances from entering the system. Such measures are by definition sustainable. A decreased use of metals and minerals from the crust of the earth will lead to a more sustainable society.

Measures used in scenarios 3 and 4 are "end-of-pipe" measures and do not reduce the total amount of substances, they merely place them somewhere else. In the infiltration scenario, a considerable amount of the studied substances end up in the soil or in the groundwater. In the pond scenario, some of the studied substances end up in the sediments. These sediments must sooner or later be taken care of, and, regardless of method, will involve a potential environmental risk.

In this study, non-structural methods like street sweeping have not been included. Although they may prove cost-effective and considerably reduce the discharges to the receiving water, they will also cause the same kind of problems as the sediments from the pond.

It is recommended that source control measures are made as a first choice and that stormwater treatment facilities are chosen only when the pollutant sources have been reduced as much as practically possible.

8. References

1. Malmqvist P-A. and Bennerstedt K. (1998) Future stormwater management in Stockholm. Case study: Hammarby Sjöstad. *6th International Symposium on Highway and Urban Pollution, Baveno, Italy.*
2. Malmqvist P-A., Larm T, Bennerstedt K, Wränghede A-K. (1999) Sources of pollutants discharging to Lake Trekanten, Stockholm, in I.B. Joliffe and J.E. Ball (eds.), *Proceedings, 8th ICUSD*, Sydney, Australia, 1999, pp. 1736-1743.
3. Engvall C. (1999) Simulations of Materials Flow in Stormwater. *Diploma Work UPTEC W 99 019*, Uppsala University School of Engineering, Uppsala, Sweden.
4. Ahlman S. (2000) SEWSYS – a modelling tool for transport and treatment processes in sewer systems developed in MATLAB/Simulink. *Diploma Work 2000:8*, Dept. of Water Environment Transport, Chalmers University of Technology, Göteborg, Sweden (In Swedish).
5. Pettersson T. J. R. (1999) Stormwater Ponds for Pollution Reduction. *PhD Thesis No 14*, Dept. of Sanitary Engineering, Chalmers University of Technology, Sweden.
6. EPA (1986) Methodology for analysis of detention basins for control of urban runoff quality, EPA440/5-87--001, U.S. Environmental Protection Agency, USA.

RISK ASSESSMENT OF STORMWATER CONTAMINANTS FOLLOWING DISCHARGE TO SOIL, GROUNDWATER OR SURFACE WATER

P.S. MIKKELSEN, A. BAUN & A. LEDIN
Environment & Resources DTU
Technical University of Denmark, Building 115, DK-2800 Kgs. Lyngby

1. Introduction

Stormwater runoff from urban areas is significantly polluted with a wide range of substances of environmental concern, and, the environmental impacts associated with wet-weather discharges have received increased attention during recent years. As a result, there are increasing pressures from society to handle stormwater runoff in ways that ensure maximum pollution reduction.

1.1. INTEGRATED URBAN STORMWATER MANAGEMENT

In order to deal with the relevant pollution problems and at the same time come up with more sustainable solutions for stormwater disposal, it is necessary to understand and manage the whole stormwater system in a holistic and integrated manner. Three sub-systems can be distinguished to underline this viewpoint:

- the *technical* sub-system,
- the *natural* sub-system, and
- the *social* sub-system.

The *technical* sub-system traditionally consists of sewer systems, detention basins and, for combined systems, wastewater treatment plants. More recently end of the pipe treatment of runoff from separate storm sewer outfalls and combined sewer overflow structures have come into focus as well as decentralised ponds and wetlands integrated into the urban landscape. Finally source control, i.e. local detention, infiltration, and re-use of stormwater has received increased attention over the past decades. The term (structural) Best Management Practices (BMPs) is commonly used in the Unites States for many of these constructive measures to handle stormwater runoff, whereas the term Sustainable Urban Drainage Systems (SUDS) is predominantly used in the United Kingdom. Non-structural BMPs are ways of managing stormwater pollution without building infrastructure, e.g. ban of products, control of building materials, and street sweeping aimed at reducing the source of pollutants that later enters the runoff water during rain.

J. Marsalek et al. (eds.), Advances in Urban Stormwater and Agricultural Runoff Source Controls, 69–80.

The *natural* sub-system is usually conceived as the "receiving" part and traditionally contains the receiving waters and their sediments. However, when infiltration is concerned, urban soils and groundwater also need to be included. Some of the structural BMPs integrate local water bodies and soil and groundwater in managing stormwater "nature's way" whereas others are based purely on constructed infrastructure. Thus, there is not always a clear distinction between technical and natural sub-systems.

The *social* sub-system is composed of the stakeholders involved with and affected by stormwater management. Traditionally, the stakeholders include policy makers and legislators as well as planners, engineers and other technicians. Although legislation is commonly implemented on the institutional level by regional, national or provincial organisations, practical projects are carried out on the local level and staff from municipalities and their consultants have major influence on decisions. Local citizens and Agenda 21 groups, however, also influence decisions as people become aware of the esthetical and ethical values of water and increasingly envisage the visibility of water, particularly rainwater, in and around urban settings as adding to the quality of life.

1.2. RISK ASSESSMENT IN A LOCAL CONTEXT

In some cases, water is an important carrier of urban planning and source control of stormwater runoff is seen as a way to cut costs by reducing the loads of stormwater discharges through sewer networks and wastewater treatment plants. In other cases, stormwater planning favours conventional centralised solutions because professionals cannot see realistic alternatives that fulfil the same basic objectives (flood control, protection of human health and pollution reduction). It is not an easy task to decide on an action or a general stormwater management policy since different interventions in the *technical* sub-system will influence different stakeholders in the *social* sub-system and impact discharges to different parts of the *natural* sub-system.

Ideally, pollutants in stormwater runoff can be discharged to either part of the natural sub-system by changing the layout of the technical sub-system. There are no (technical) standard solutions that solve problems equally well at all locations, and it cannot be known in advance whether it is better from an environmental point of view to discharge stormwater to surface water recipients or to the soil-groundwater system. For many reasons stormwater management decisions need to consider the local context, i.e. the local natural and technical preconditions, but also the local political and societal organisation.

Clearly, methodologies and tools are needed for assessing the environmental risks associated with discharge of stormwater pollutants to the environment. Such methodologies should acknowledge that stormwater projects cover a wide range of technologies and spatial scales, that they commonly have limited budgets, and that they are dealt with by non-experts in environmental risk assessment. This paper reviews some concepts used within risk assessment of chemical substances and seeks to make a course for further developments related to risk assessments of stormwater contaminants

2. Evidence of pollution with stormwater contaminants

Historically, there has been a lot of research on chemicals in stormwater runoff. Currently in Denmark, there is increased focus on risk assessment related to discharge to surface waters. Groundwater quality is also sometimes considered – but soil contamination is never thought of. Examples based on Swiss conditions of pollution of the different environmental compartments are given below. Heavy metals and polyaromatic hydrocarbons, which are both well-known groups of stormwater contaminants, are used in these examples.

Table 1 gives typical concentration levels in road runoff of copper (Cu), zinc (Zn), cadmium (Cd) and lead (Pb) from a Swiss literature survey [1]. Also shown are eco-toxic concentrations calculated by applying extrapolation factors to toxicity data and proposed Swiss river water quality criteria [2]. It appears that the concentrations of these heavy metals exceed the defined quality limits and ecological damage is therefore expected when discharging untreated road runoff to surface waters. This conclusion was recently confirmed by testing the toxicity of urban runoff directly [3].

TABLE 1. Comparison of typical metal concentrations in road runoff with eco-toxic values, proposed Swiss river quality criteria and Swiss drinking water quality criteria. All concentrations are in µg/l.

| Concentration in road runoff [1] | | Environmental quality limits | | | |
| | | River water [2] | | Drinking water [4] | |
		Eco-toxicity	Proposed std.	Target	Threshold
Cu	40 – 380	0.05 – 2	2	50	1500
Zn	166 – 1950	0.5 – 2	5	100	5000
Cd	1.4 – 20	0.005 – 0.5	0.05	0.5	5
Pb	100 – 980	3 – 10	1	1	50

The Swiss drinking water criteria [4] are also shown in Table 1. The target values represent average concentration levels in groundwater uninfluenced by anthropogenic activity and the threshold values represent the accepted quality of drinking water when it leaves the water works. The standard of reference depends strongly on the context, i.e. all four metals exceed the target values whereas only Cd and Pb seem to exceed the threshold values.

Figure 1 shows results from a field investigation of infiltration shafts located in Basel, Switzerland where road runoff had infiltrated for more than thirty years [5]. A few decimetres of pebbles were found at the bottom of the shafts on top of the original coarse gravel deposits. Runoff materials (road sediment mixed with organic material from nearby trees and bushes) had accumulated above and between the pebbles and penetrated 70 cm into the gravel. The measured solid phase concentrations of Zn and PAH were closely connected with the sludge layer. Zn exceeded the Swiss soil quality standard of 200 mg Zn kg^{-1} [6] and ΣPAH (sum of 16 individual compounds defined by the US EPA) exceeded the A-value from the Dutch system of reference values for soil quality [7]. For comparison, the concentrations beneath the sludge layer and in a control profile not influenced by infiltration were insignificant.

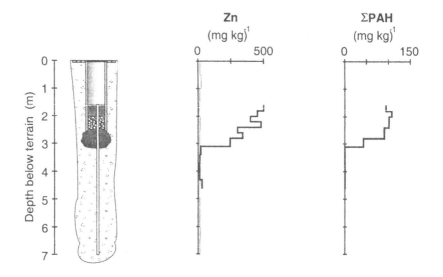

Figure 1. Depth distribution of zinc (Zn) and polyaromatic hydrocarbons (ΣPAH) through an infiltration shaft and into the underlying gravel. The dotted line shows concentrations in a control profile uninfluenced by infiltration.

As previously mentioned, contaminants in stormwater runoff can be discharged to either part of the *natural* sub-system (i.e. surface water/sediments, soils or groundwater) by changing the layout of the *technical* sub-system. As an example, Boller made mass balances for water-carried copper (Cu) in the Swiss town St. Gallen [8]. Copper is available at high concentrations in stormwater runoff in Switzerland due to the extensive use of copper for gutters and roof materials. Three idealised layouts of the urban stormwater drainage system were considered: combined sewers, separate wastewater and stormwater sewers, and separate wastewater sewers in combination with stormwater infiltration. About 2/3 of the total copper load originates from stormwater runoff. With a combined sewer system, 71% accumulates in sewage sludge while 24% goes to receiving water and the remaining 5% classifies as urban diffuse pollution. Separation of the sewer system turns the picture around by leading 75% of the total copper load into the surface water recipient; only a small part thereof comes from illicit cross connections between wastewater and stormwater pipes. The copper mass flow that leads to well-known contamination of receiving waters and sediments is almost completely redirected into urban soil and groundwater when stormwater infiltration is used.

Although the list of physical, chemical and biological constituents and parameters that have been observed in stormwater runoff is extremely long, focus has mostly been on selected key-constituents such as settleable particles, chloride (from road salting), organic matter (BOD/COD), heavy metals, polyaromatic hydrocarbons (PAH), E-coli, etc. Pesticides and other organic chemicals related to the use of products and erosion of the built environment have only recently come into focus. Bioaccumulating compounds (as e.g. heavy metals and PAH in the examples above) have received the most attention because they end up in parts of the environment directly exposed to humans (i.e. sediments and surface soils). The substances of greatest environmental risks in relation to contamination surface water via direct discharge or groundwater via infiltration of stormwater runoff (i.e. highly soluble compounds) are those that have been studied the least. In a recent Swiss investigation the pesticide Atrazine was found in roof runoff at median and maximum concentrations of 0.033 µg/l and 0.903 µg/l respectively, which is quite close to, and above the drinking water quality limit of 0.1 µg/l [9]. Furthermore, Atrazine penetrated through an infiltration system without seemingly being reduced due to degradation or sorption.

The 'traditional' list of key-substances is not exhaustive; 215 individual chemical stormwater constituents were identified in a brief literature review [10] that covered the following groups as well as some organic contaminants that did not fit into the groups:

- Heavy metals
- Other inorganic compounds
- PAH
- Phenols and cresols
- PCB and pesticides
- Plasticisers, halogenated alifatics, monocyclic aromatics

Society produces thousands of new chemicals every year, and most of them will find their way into stormwater runoff and eventually be identified as the analytical detection limits are continuously improved. The international literature reports on an increasing number of investigations that focus on chemicals in stormwater runoff and their related toxicity. This development calls for proper risk assessments that take the properties of each compound into account when assessing the risk associated with their discharge to the surface water–sediment or the soil–groundwater system. To make a step forward in this direction, it is worthwhile to review some of the generic concepts and principles established within environmental chemistry and eco-toxicology for risk assessment of chemicals.

3. Risk assessment of chemical substances

3.1. TERMINOLOGY AND PROCEDURES

Figure 2 gives an overview of the terminology used in risk assessment of new and existing chemicals according to the technical guidance document (TGD) for risk assessment of chemicals in the EU [11]. Risk assessment of chemicals is composed of four elements: hazard identification, hazard assessment, risk characterization and risk management. The cycle to the left illustrates that risk assessment is not a linear process but that iterations in the four steps are necessary depending on the problem and the available data.

Hazard identification serves to map the inherent properties of chemicals by collecting and comparing relevant data on e.g. physical state, volatility, mobility,

74

degradability, bioaccumulation and toxicity. It is thus the basic step of procuring data that is needed when proceeding with the following steps.

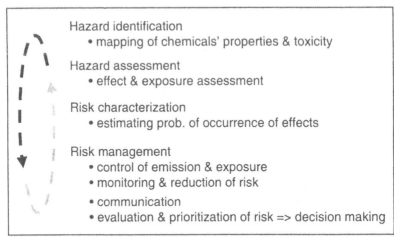

Figure 2. Terminology used in risk assessment of chemical substances [11]

Hazard assessment is divided between exposure assessment and effect assessment. Comprehensive model systems have been developed to assess the distribution of contaminants in the environment (soil, water, air) and in tissue (animals and humans). The EUSES model developed with support from the European Union is illustrated in Figure 3 as it is implemented in the TGD [11]. The complexity of this model is illustrated by the fact that input is entered via more than 100 menus and that it has almost 500 parameters of which about 40 need to be defined for each simulation. The output is given as *predicted environmental concentrations* (PECs) that are estimates of the *exposure* of the environment with individual chemical compounds.

The *effects* (both acute and chronic) associated with the presence of each substance in the environment and tissue needs to be assessed by assessing available data on toxicity to humans and ecosystems (i.e. dose-response relationships, EC50, LC50 etc.). In this manner, *predicted no-effect concentrations* (PNECs) can be calculated and compared with the predicted environmental concentrations (PECs). If the requirement shown in equation (1) is fulfilled, no environmental problem is said to be present; however, if it is not, then there is a problem that should be dealt with somehow.

$$\frac{PEC}{PNEC} < 1 \quad \Leftrightarrow \quad PEC < PNEC \tag{1}$$

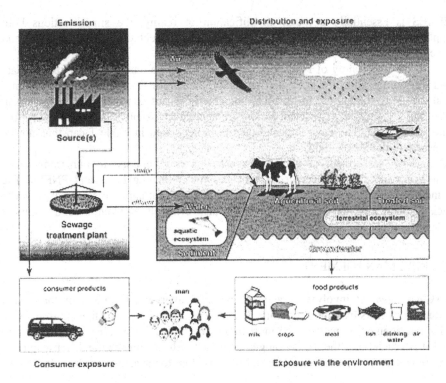

Figure 3. Outline of the exposure model EUSES [11, 12]

3.2. RISK AND UNCERTAINTY

After PECs and PNECs are estimated the *risk characterization* takes place (see Figure 2). The potential negative effects are evaluated and, if possible, the probability of effects occurring is estimated. This conforms with the basic definition of risk being equal to the product of probability and consequence:

$$\text{Risk} = \text{probability} \times \text{consequence} \tag{2}$$

Theoretically, risk is only high if it is probable (or likely) that an effect with consequence occurs. A rare but severe effect may ideally have the same risk as a frequent but less severe effect, implying that different risks can be compared and evaluated.

The PECs and PNECs estimated as described above are very uncertain. The uncertainty of PECs stems from the difficulty with identifying and describing the processes responsible for the distribution of each contaminant and the inability to quantify the parameters controlling these processes. Even a complex model like

EUSES is essentially a rough simplification of reality, the simplifications both applying to the description of processes and the representation of the technical and natural sub-systems.

The uncertainty of PNECs can also be very large due to lack of toxicity data and difficulties with, for example, translating data from laboratory experiments into realistic field conditions. In practice, uncertainty is not considered as suggested in equation (2). Instead, assessment factors are applied to toxicity data to account for the uncertainty [11]. The environmental quality limits stated in Table 1 can thus not be interpreted directly as PNECs.

Risk assessment of chemical substances can be done in numerous ways, employing more or less detailed approaches for estimating PECs and PNECs. The principles described above are quite detailed and will not apply to some cases due to lack of data. Less detailed approaches are also available employing semi-empirical rules-of-thumb, see, for example, [13].

3.3. RISK MANAGEMENT

Risk management involves a range of possible interventions, i.e. monitoring and reduction of risks by control of emissions and the exposure of sensitive environments. Risk management is perhaps the most important part of a risk assessment procedure and it is crucial that the methodologies used are well documented and transparent to ensure that the risk perceived by the public is as close to the estimated risk as possible.

4. Discussion

Ideally, the principles explained in the previous section can be used directly when assessing the risk associated with discharge of stormwater contaminants. However, there are a number of problems as discussed below.

4.1. PROBLEMS WITH RISK ASSESSMENT OF STORMWATER CONTAMINANTS

First, some problems exist in relation to hazard identification of stormwater contaminants. Some of the most important problems are as follows.
- The number of chemical compounds observed in stormwater runoff is very large and is increasing as analytical methods are continuously improved and more field studies are conducted due to increased attention from society.
- The measured concentration levels vary over several orders of magnitude due to the intermittent character of the runoff process, the plenitude of pollutant sources and the fact that investigations tend to emphasize the extremes (high concentrations).
- Basic physical/chemical properties are available for only a few substances, and the toxicity of stormwater runoff, a complex mixture of substances in the presence of particles and organic matter, has only been studied a little (and very recently).

Effect assessment of stormwater contaminants can be conducted by comparisons with defined environmental quality limits for:

- Soil – aimed at preserving soil organisms and fertility for plant production, and protecting children from poisoning via soil ingestion.
- Groundwater – aimed at preserving raw water in a quality suitable for human consumption.
- Surface water – aimed at preventing acute toxicity to aquatic organisms and plants.
- Sediments – aimed at preventing chronic toxicity to benthic organisms and plants.

However, such quality limits are available only for a few 'classical' contaminants.

It is interesting to note that although wastewater treatment plants are represented in EUSES, stormwater systems are not represented at all in relation to risk assessment of chemicals in the EU. The methodology is directed towards assessment of industrial chemical products and only the major compartments and pathways are included. This means that the EUSES model cannot be used for exposure assessment in relation to risk assessment of stormwater contaminants.

Instead, models need to be developed as a supplement to, or in parallel to, the EUSES system. These models should describe the physical stormwater systems in great enough detail to facilitate comparing different stormwater BMPs; however the behaviour of individual compounds should be described using the same chemodynamic principles that are used in EUSES.

4.2. AN EXAMPLE OF HAZARD ASSESSMENT: STORMWATER INFILTRATION

A simplified hazard assessment that applies to stormwater management is illustrated below. The example is about soil contamination with Zn due to stormwater infiltration in a swale with 20 cm of organic soil with pH~7 at the top. Zn is rather strongly sorbed to the solid phase under these circumstances and it can be assumed that the Zn carried with the incoming water accumulates in the soil layer. Thus, the soil concentration (PEC) can be calculated as

$$C_s = C_w \frac{A_r}{A_i} \frac{MAR}{d\rho} t \quad \text{(mg/l)} \tag{3}$$

where C_w is the concentration of Zn in runoff water (mg/l), A_r/A_i is the ratio between the (impermeable) runoff area (-), MAR is the mean annual rainfall (0.5 m/year), d is the thickness of the soil layer (0.2 m), ρ is the bulk density of the soil (kg/l) and t is the time elapsed (years).

Mean runoff concentrations of Zn from different surface areas are used as input to the calculation as illustrated in Figure 4 [13]. The relatively narrow concentration limits (0.3-0.5 mg Zn/l) for separate storm sewers in mixed land-use areas is due to the large catchment scale [14]. This is reflected in distinct confidence areas for A_r/A_i=5 and 25. The concentrations vary much more for smaller catchment areas (highways or

78

roofs) making it more difficult to distinguish between soil pollution loading for different values of A_r/A_i. It may take anywhere between 10 years and several hundred years to reach soil concentrations of 200 mg Zn/kg soil, corresponding to the Swiss soil quality criterion. Such calculations are naturally very approximate and apply only for substances that strongly accumulate in the surface layers. However, the calculations clearly demonstrate the uncertainty related to predicting the mass load in planning situations where no measurements of runoff quality are available. This large uncertainty implies that exposure models should not be too detailed.

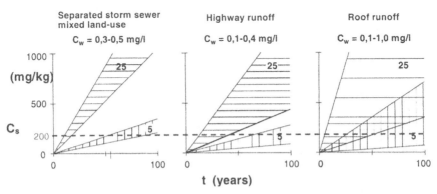

Figure 4. Accumulation of Zn in the top 20 cm of an infiltration swale by a simple mass balance approach according to equation (3). The dotted line at C_s=200 mg/kg illustrates the Swiss soil quality criterion for Zn [6].

4.3. OPTIONS FOR ACTION

When discharging contaminated stormwater runoff there are (only) three options for action.
• Contaminants can be removed at the source, e.g. by controlling the use of building materials giving rise to erosion and dissolution of contaminants, or by cleaning urban surfaces regularly (street sweeping).
• Contaminants can be immobilised or degraded in stormwater BMPs designed for the specific purpose of pollution reduction.
• Contaminants can be discharged in the least harmful manner.

To analyse these options, models are needed that describe the sources and fluxes of stormwater contaminants, including their fate during passage through stormwater BMPs and final disposal in the environment. In addition, risk assessment methodologies are needed that facilitate comparing scenarios based on different BMPs. Model based risk assessment using a probabilistic approach based on equation (2) allows for combining risk assessment with traditional cost-benefit analyses [15]. It may even help analysing whether it is better from an environmental point of view to discharge stormwater contaminants to the soil-groundwater system or the surface water-sediment system.

5. Conclusions

The basic challenge for future risk assessment of stormwater contaminants is to ensure that the methodologies used account for information about the local context and produce information that feeds into local decision making processes. It is essential that sound scientific knowledge about the chemodynamics and toxicity of chemical substances is considered to ensure a minimum level of transparency, and learning from principles developed for risk assessment of chemical substances best ensures this.

In the future we will need tools of varying complexity that can be applied on various spatial scales by acknowledging that stormwater projects are often small, have limited budgets and are dealt with by non-experts in chemical risk assessment. Some useful tools would be as follows:

- an identification list of those stormwater constituents which can be specifically classified as contaminants following discharge to surface water, groundwater and/or soil;
- a database of physical/chemical properties of these contaminants, their environmental fate and effects, their predicted no-effect concentrations, when available, and their treatability (possible removal in various BMPs);
- a tool for estimating concentrations of these stormwater contaminants in a site-specific context; and
- a screening tool for identifying in a site-specific context which stormwater contaminants are critical (priority pollutants) following discharge to surface waters, groundwater and/or soils.

6. References

1. Häfliger, M. and Boller, M. (1994) Verbleib von Schwermetallen bei unterschiedlichen Konzepten der Siedlungsentwässerung (in German). Swiss Federal Institute for Environmental Science and Technology (internal report).
2. Behra, R., Genoni, G.P., and Sigg, L. (1994) Festlegung der Qualitätsziele für Metalle unde Metalloide in Fliesgewässern - Wissenschaftliche Grundlagen (in German). *GWA* 73(12), 942-951.
3. Marsalek, J., Rochfort, Q., Brownlee, B., Mayer, T., and Servos, M. (1999) An exploratory study of urban runoff toxicity, *Water Science and Technology* 39, 33-39.
4. Schweizerisches Lebensmittelhandbuch, Kapitel 27A, Trinkwasser (in German), 1988.
5. Mikkelsen, P.S., Häfliger, M., Ochs, M., Tjell, J.C., Jacobsen, P., and Boller, M. (1996) Experimental assessment of soil and groundwater contamination from two old infiltration systems for road run-off in Switzerland, *The Science of the Total Environment* 189/190, 341-347.
6. Verordnung vom 9. Juni 1986 über Schadstoffe im Boden, SR 814.12; einschliesslich Erläuterungen des BUWAL vom Juni 1987.
7. Moen, J.E.T., Cornet, J.P., and Ewers, C.W.A. (1985) Soil protection and remedial actions: Criteria for decision making and standardization requirements, in J.W. Assink and W.J. van den Brink (Eds.), *Contaminated soil '86. First Int. TNO Conf. on Contaminated Soil, Utrecht, The Netherlands*, 441-448.
8. Boller, M. (1997) Tracking heavy metals reveals sustainability deficits of urban drainage systems. *Water Science and Technology* 35(9), 77-87.

9. Bucheli, T.D., Müller, S.R., Heberle, S., and Schwarzenbach, R.P (1998) Occurrence and behaviour of pesticides in rainwater, roof runoff, and artificial stormwater infiltration, *Environmental Science and Technology* **32**, 3457-3464.

10. Lindstrøm, M. and Nielsen, M.H. (1999) Local disposal of stormwater – Risk assessment of groundwater pollution (M.Sc. thesis, in Danish). Department of Environmental Science and Engineering, Technical University of Denmark.

11. European Commission. (1996) Technical Guidance Documents in Support of the Commission Directive 93/67/EEC on Risk Assessment for New Notified and Commission Regulation (EC) No1488/94 on Risk Assessment for Existing Substances. Bruxelles, Belgium.

12. van Leeuwen, C.J. and Hermens, J.L.M. (Eds.) (1996) *Risk assessment of chemicals – an introduction*, Kluwer Academic Publishers, Dordrecht, The Netherlands.

13. Mikkelsen, P.S. (1995) Hydrological and pollutional aspects of urban stormwater infiltration (Ph.D. thesis), Department of Environmental Science and Engineering, Technical University of Denmark.

14. Mikkelsen, P.S., Weyer, G., Berry, C., Walden, Y., Colandini, V., Poulsen, S., Grotehusmann, D., and Rohlfing, R. (1994) Pollution from urban stormwater infiltration, *Water Science and Technology* **29(1-2)**, 293-302.

15. Hauger, M.B., Rauch, W., Linde, J.J., and Mikkelsen, P.S. (2001) Cost-benefit-risk – A concept for management of integrated urban wastewater systems?, Proc. 2nd Int. Conf. on Interactions Between Sewers, Treatment Plants and Receiving Waters in Urban Areas – INTERUBA II, Lisbon/Portugal, February 19-22, 8 p.

CASE STUDIES OF LOCAL STRATEGIES FOR CONTROL OF NON-POINT SOURCE POLLUTION IN COLORADO (USA)

T. A. EARLES, J.E. JONES, & W.F. LORENZ
Wright Water Engineers, Inc.
2490 West 26th Avenue, Suite 100A
Denver, Colorado 80211, USA

1. Introduction

In the 1998 National Water Quality Inventory report to Congress, the State of Colorado reported that eighty-nine percent of the State's 171,000 river kilometres and ninety-one percent of the 59,000 lake hectares in the State have "good" water quality, fully supporting designated uses [1]. Maintenance of this high level of water quality is important from the standpoints of water supply, protection of aquatic ecosystems, and sustenance of industries that rely on the pristine environmental conditions for which Colorado is known. Primary sources of non-point source (NPS) pollution in Colorado include agriculture, mining, construction, and urban runoff, with much of the lake impairment in the State attributable to the latter two sources [1]. NPS pollutant sources from construction and urban runoff have the potential to increase dramatically in coming years due to population growth and associated urbanisation; population is projected to be nearly two times 1990 levels by 2025 for the Denver metropolitan area and for the State as a whole [2].

While non-point source discharges are regulated on the Federal level through the National Pollutant Discharge Elimination System (NPDES) of the Clean Water Act (CWA) and on the State level through the Colorado Discharge Permit System (CDPS), local agencies are taking responsibility for water quality protection through local initiatives, ordinances, and development requirements. These local controls typically address many issues including:

- Wetlands protection.
- Preservation of undisturbed buffer zones adjacent to waterbodies and wetlands.
- Stream habitat preservation and restoration.
- Stormwater quality management.
- Land use and zoning restrictions, including the creation of "overlay" districts.
- Source water protection requirements, often related to public water supplies.
- Limitations on the amount of imperviousness created in a watershed.
- Special erosion and sediment control requirements.
- Threatened, endangered, rare, and sensitive species.
- Preservation of groundwater supplies.
- Minimum streamflow.

81

J. Marsalek et al. (eds.), Advances in Urban Stormwater and Agricultural Runoff Source Controls, 81–93.

This paper presents case studies of local initiatives, both in the Denver metropolitan area and in the mountain region, to illustrate strategies for controlling sources of NPS pollutants in the rapidly growing urban areas of Colorado as well as for minimising impacts from development in sensitive areas. Case studies include local ordinances addressing water quality and wetlands protection as well as an agreement imposing specific water quality requirements, including numeric limits, on stormwater discharges from residential development. Case studies of local ordinances include:

- Cherry Creek Reservoir Watershed Stormwater Quality Model Ordinance [3].
- Town of Silverthorne Waterbody, Wetland and Riparian Protection Regulations [4].

In addition to these case studies, the Grant Ranch stormwater quality management and monitoring program [5] is presented as a case study of a project-specific agreement for stormwater management and source control to illustrate strict water quality protection requirements driven by local concern.

2. Source Control Ordinances

2.1 CHERRY CREEK RESERVOIR WATERSHED STORMWATER QUALITY MODEL ORDINANCE

The Cherry Creek Reservoir is located in the southeastern portion of the Denver metropolitan area in Arapahoe County. The reservoir has a surface area of approximately 340 hectares and a drainage basin of approximately 1000 square kilometres. The drainage basin is largely composed of rapidly urbanising portions of Arapahoe and Douglas counties and contains a relatively small portion of El Paso County at the headwaters of the creek. Land uses in the drainage basin include considerable areas of commercial development, numerous large office parks, and the communities of Parker and Franktown. The Cherry Creek Basin is shown in Figure 1. The mean depth of the reservoir is approximately 3 metres with a maximum depth of 6 metres in the vicinity of the outlet works. The Cherry Creek Reservoir is owned and operated by the U.S. Army Corps of Engineers and is surrounded by Cherry Creek State Park, an approximately 1500 hectare multi-use recreational area. Originally, the Cherry Creek Reservoir was constructed for flood control purposes in 1950; however, once the reservoir was filled, the potential for multiple uses became evident [6].

Classified uses of the reservoir include [7]:
- Aquatic Life 1 - Warm waters that (1) currently are capable of sustaining a wide variety of warm water biota, including sensitive species, or (2) could sustain such biota but for correctable water quality conditions. Waters shall be considered capable of sustaining such biota where physical habitat, water flows or levels, and water quality conditions result in no substantial impairment of the abundance and diversity of specifies [8].

- Recreation 1 - waters that are suitable or intended to become suitable for recreational activities in or on the water when the ingestion of small quantities of water is likely to occur. Such waters include but are not limited to those used for swimming, rafting, kayaking and water-skiing [8].
- Agriculture - waters that are suitable or intended to become suitable for irrigation of crops usually grown in Colorado and which are not hazardous as drinking water for livestock [8].
- Water Supply - waters that are suitable or intended to become suitable for potable water supplies. After receiving standard treatment (defined as coagulation, flocculation, sedimentation, filtration, and disinfection with chlorine or its equivalent) these waters will meet Colorado drinking water regulations and any revisions, amendments, or supplements thereto [8].

The 1982 Clean Lakes Study identified potential for negative impacts to the reservoir's beneficial uses from eutrophication. Assessments of the reservoir's trophic state from 1992 through 1996 using Carlson's Trophic State Index (TSI) based on mean values of Secchi depth, chlorophyll a, and total phosphorus from April through October indicated that the reservoir was eutrophic [6]. As a result, phosphorus is a pollutant of primary concern in the Cherry Creek Basin.

The Cherry Creek Basin Water Quality Authority ("Authority"), a quasi-municipal corporation and political sub-division of the State established under the statutory authority of the state, is responsible for water quality in the Cherry Creek Basin. The Cherry Creek Reservoir Watershed Stormwater Quality Model Ordinance ("Model Ordinance") was developed in 1999 by the Authority for the following stated purpose:

To provide substantive requirements to control the quality of stormwater runoff in the Cherry Creek Basin from private and public property and to reduce the loads of contaminants reaching Cherry Creek and the Cherry Creek Reservoir in furtherance of health, safety, and general welfare in the Cherry Creek Basin. This model ordinance applies to both construction and post-construction development. The recommended stormwater quality Best Management Practice (BMP) requirements set forth in this model ordinance...are necessary to reduce and maintain non-point source phosphorus loads below their load allocation, in accordance with Total Maximum Daily Load ("TMDL") set forth in the Cherry Creek Control Regulation...and the water quality requirements delineated in the Cherry Creek Basin Water Quality Master Plan...

84

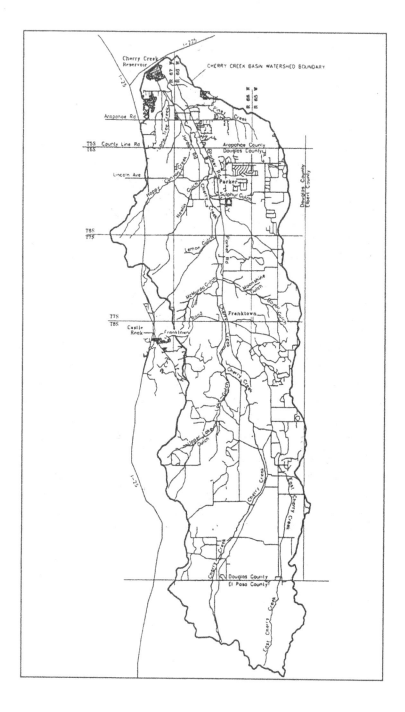

Figure 1. Cherry Creek Basin

The Model Ordinance seeks to control NPS pollution in the basin through regulation of land disturbing activities and requires that municipalities, counties, and other governmental entities with jurisdiction over land use in the basin adopt the BMPs and policies set forth in the Model Ordinance. The Model ordinance requires adoption of temporary (construction) and permanent BMPs for land disturbances including grading, cutting and filling, building, paving, and other anthropogenic activities that change the natural cover or topography of the land surface. Under the Model Ordinance, the owner of a development project is required to submit an application and receive approval for any land disturbing activities. The owner must prepare a site-specific plan for stormwater quality control, meeting the BMP requirements specified in the Model Ordinance. The Model Ordinance specifies requirements for construction BMPs and permanent BMPs and additionally contains specific requirements for industrial and agricultural activities and land disturbances occurring in stream preservation areas. Stream preservation areas include the reservoir, Cherry Creek State Park, drainage and discharge to the park within 30 metres of the park boundary, lands within the 100-year flood plain of Cherry Creek, and land overlying the Cherry Creek alluvium.

Individual home construction is exempted from the application and approval requirements of the Model Ordinance provided that the land disturbance is less than 0.4 hectares, the home is not a part of a larger development by the same owner, and sediment entrapment BMPs including silt fence, filter strips, sediment basins, and/or straw bale barriers are implemented on-site.

A major component of the ordinance is the requirement of a Construction BMP (CBMP) Plan. The CBMP Plan must include:

- A narrative description of the construction project.
- Construction schedule, including implementation of BMPs.
- Detailed descriptions and locations of BMPs and justification that planned BMPs will satisfy Model Ordinance criteria.
- Site map and drawings showing the area of disturbance; existing and proposed topography; areas of excavation, grading, and fill; and locations of proposed BMPs.
- Characterisation of site geology and soils.
- Inspection and maintenance procedures, including access considerations and site-specific inspection log. Inspection is required after installation of any construction BMPs, after any runoff event causing erosion, and at least once a month. Inspections must be documented, and inspection logs must be maintained on-site.

Land disturbing activities are to be accomplished in a way that satisfies requirements of six categories of pre- and during-construction BMPs specified in the Model Ordinance. Pre-construction BMPs must be implemented before any land disturbing activities begin, while during-construction BMPs for disturbed areas must be implemented within 14 days of soil disturbance. Categories of BMPs that must be included in the CBMP Plan include:

- Category 1: Phasing of Construction (pre-construction) - construction activities must be scheduled to minimise the amount of soil that is exposed at a given time, and areas of disturbance larger than 40 acres may not remain exposed for longer than 30 days.

- Category 2: Reduction of Runoff Flows to Non-erosive Velocities (pre-construction) - runoff flows up-gradient of disturbed areas must be diverted around disturbed areas, and runoff velocities across disturbed areas must be limited to less than 0.6 metres per second (non-erosive velocity) over bare areas. BMPs include swales, diversion dikes, terracing and contouring, slope drains, and/or check dams.

- Category 3: Protection of Drainageways from Erosion and Sediment Damages (pre-construction) - to prevent sediment from disturbed areas from entering drainageways and to keep runoff flows from eroding drainageways, BMPs including waterway crossing protection, inlet protection, outlet protection, and/or temporary diversions must be applied.

- Category 4: On-site Retention of Sediment (pre-construction) - sediment entrapment measures including silt fence, filter strips, sediment basins, and/or straw bale barriers are required to prevent accelerated soil erosion, impede sediment movement, and reduce off-site transport of sediment. Vehicle tracking control must also be provided, including plans for daily maintenance.

- Category 5: Stabilisation of Exposed Soils (during-construction) - areas that will remain exposed and/or inactive for more than fourteen days must be stabilised by mulching; seeding; tackifer application; use of erosion control mats, blankets, and nets; and/or surface roughening.

- Category 6: Revegetation of Disturbed Areas (during-construction) - temporary and permanent revegetation of disturbed areas are required within 14 days of temporary or permanent cessation of construction. Temporary revegetation is required for areas that will be exposed during any growing season prior to completion of construction, and permanent revegetation is required for disturbed areas that will be exposed for two-or more years or indefinitely.

In addition to a CBMP Plan, the Model Ordinance requires the development of a Permanent BMP (PBMP) Plan to be submitted in two phases: (1) a preliminary phase defining the nature of the development and proposed permanent BMPs and (2) a final phase providing more in-depth discussion of concepts identified in the preliminary PBMP Plan and design details for BMPs. Requirements of the PBMP Plan include:

- Narrative description of development project and map showing location of proposed development.

- Map showing drainage area; area of disturbance; existing and proposed topography; areas of excavation, grading, and fill; and locations of proposed BMPs.

- Characterisation of site geology and soils.

- Discussion of relationship to regional and basin drainage and stormwater quality plans.

- Detailed description, design criteria, location, and operation and maintenance requirements for BMPs including description of how BMPs satisfy water quality capture volume (WQCV), pollutant removal, and maintenance requirements of the Model Ordinance.
- Schematics of proposed BMPs (preliminary PBMP Plan).
- Maintenance and inspection protocols for BMPs and commitments for maintenance from responsible agencies/owners, including dedication of any necessary easements for maintenance access.
- Detailed design drawings including size, location, specifications, and technical details for all permanent BMPs (final PBMP Plan).
- Schedule for construction and operation of BMPs (final PBMP Plan).

Permanent BMPs are required to provide a WQCV designed to capture and treat the 80th percentile runoff event in accordance with the Denver Urban Drainage and Flood Control District's Drainage Criteria Manual, Volume III [9]. BMPs in the model ordinance consist of detention measures including:

- Extended dry ponds - 40-hour drain time for WQCV with additional 20% of WQCV provided for sediment accumulation.
- Wet ponds - 12-hour drain time for WQCV.
- Constructed wetlands basins - 24-hour drain time for WQCV.
- Porous pavement detention - 6-hour drain time for WQCV plus 2-inch surcharge storage volume for WQCV from adjacent areas.
- Porous landscape detention - 12-hour drain time for WQCV with additional 20% of WQCV provided for sediment accumulation.
- Sand filter detention - 40-hour drain time for WQCV.

The use of BMPs in combination is encouraged, as is the use of regional facilities serving multiple development projects. Recommended combinations of BMPs include the use of detention measures (described above) in conjunction with wetland channels and grass swales. Design depth requirements for channels are dictated by the two-year runoff event and the degree of detention provided by the channel varies from 6- to 12-hours depending on the detention provided up-gradient of the channel.

While the Model Ordinance contains provisions that are, in many respects, congruous with requirements for a stormwater discharge permit under the State's CDPS system, the Cherry Creek Basin Model Ordinance provides a more refined level of regulation for several reasons. The Model Ordinance regulates a wider spectrum of development projects by excepting only individual home construction projects disturbing less than 0.4 hectares. CDPS stormwater permits are currently required for land disturbing activities disturbing 2 or more hectares (under Phase II of NPDES, this threshold will be lowered to 0.4 hectares, but development projects below this threshold will not require a permit). In addition, the Model Ordinance provides a greater degree of regulation by requiring review of CBMP Plans. A CDPS stormwater permit requires development and implementation of a Stormwater Management Plan (SWMP); however, there is no process for review. Finally, the Model Ordinance provides a greater level of detail than regulations on the State level by specifying design criteria for BMPs.

2.2 TOWN OF SILVERTHORNE WATERBODY, WETLAND AND RIPARIAN PROTECTION REGULATIONS

The Town of Silverthorne is located in Summit County, Colorado at an altitude of approximately 9,000 feet. The Town is experiencing rapid growth—population has nearly doubled since 1990, and the prevalence of second homes and vacation rentals in this resort community results in considerable construction activity in the Town. There are many wetlands and streams within Town boundaries, including "gold medal" trout water (the Blue River) running through the centre of the Town.

Although Section 404 of the CWA regulates dredging and filling of waters of the United States, including wetlands, the Town Council concluded that the Federal wetland permitting program did not provide adequate protection for local wetlands, lakes, and streams, primarily because the Section 404 regulations did not directly address the need for buffer zones—setbacks from the edges of wetlands and waterbodies. As a result of the local desire for a higher level of wetland and waterbody protection and provision of buffer zones, the Town of Silverthorne enacted Waterbody, Wetland and Riparian Protection Regulations, Ordinance 1999-1 [4] in 1999.

The intent of these regulations is:

... to protect the vital beneficial functions and values of wetlands and water areas within Silverthorne. This is to be accomplished by requiring that a permit be obtained for development activities in wetlands and water areas, and associated buffer areas, and that, as a part of this permitting process, the Town will review disturbance permits and mitigation plans. A critical element of this process is the determination of the buffer area boundary, which will vary from 25 feet [7.6 metres] to 125 feet [38 metres] depending upon the presence of site-specific features and the use of best management practices. It is intended that the buffer width will equal what is necessary to protect the wetlands and water areas from significant adverse impact arising form activities within the buffer and that applicants will be encouraged to reduce the width of the buffer through appropriate best management practices. It is anticipated that where an applicant includes best management practices, the width of the buffer will be decreased below the maximum, and that where these practices will fully mitigate the impact of development within the maximum buffer upon the wetlands or water area, the buffer will be reduced to the minimum of 25 feet [7.6 metres].

Since the Silverthorne regulations specify a buffer between areas of development and wetlands, waterbodies, and riparian areas that can vary in width from 7.6 metres to 38 metres, determining the required width of the outer buffer zone for a specific development project is critical. To accomplish this, the Silverthorne regulations identify twelve criteria for determination of the width of the variable-width buffer zone. These criteria account for well-documented functions of wetland/waterbody buffer zones including reduction of sediment and other pollutant loads to wetlands and waterbodies, attenuation of runoff rates and volumes, provision of wildlife habitat and diversity of vegetation, groundwater recharge, and soil stabilisation. Where BMPs are utilised, the width of the outer buffer may be reduced. To determine the width of the outer buffer zone, the following twelve criteria are applied [4]:

- Riparian Area Boundaries - the outer buffer zone must be coincident with the outermost boundary of riparian areas.
- Threatened and Endangered Species - the outer buffer zone will be established at least 7.6 metres to 30 metres from the outermost boundary of occupied functional habitat for plant and animal species listed by the State or Federal government as threatened or endangered.
- Wildlife Mitigation Corridor - in areas where there is a demonstrated wildlife migration corridor (as identified by the Colorado Division of Wildlife) an outer buffer zone of 7.6 metres to 30 metres will be required.
- Fens - where fens are present, an outer buffer of 30 metres will be required.
- 100-year Flood Plain - the outer buffer boundary must be coincident with the outermost boundary of the 100-year flood plain.
- Steep Slopes - where there is a slope of 20 percent or more within 38 metres of and draining to wetlands or waterbodies, an outer buffer width of 30 metres will be required; where is a slope of 30 percent or more with a vertical height that exceeds 3 metres within 38 metres of and draining to wetlands or waterbodies, an outer buffer width will be coincident with the area of the slope within 38 metres of the wetland or waterbody.
- Erodible Soils - where there are soils with a Natural Resource Conservation Service (NRCS) "k" factor of 0.25 or greater, the outer buffer zone must be coincident with the area of such soils within 38 metres of the wetland or water body.
- Unstable Stream Banks - the outer buffer zone must be at least 7.6 metres when horizontal or vertical degradation of stream banks exceeds natural levels.
- Hazardous Materials - a minimum outer buffer of 30 metres is required when proposed use of property presents a special hazard to water quality resulting from storage, handling, or use of hazardous or toxic materials, chemical fertilisers, or pesticides (residential uses are excepted).
- Stormwater Permit - when the proposed land use of the property requires a commercial or industrial CDPS permit, a minimum outer buffer of 30 metres is required.
- Impervious Area - an outer buffer of 30 metres is required when any one-acre area within the potential buffer area will have an imperviousness of 40 percent or more.
- Poor Vegetative Cover - an outer buffer of 15 metres is required if vegetative cover is less than 30 percent for any 0.5-acre area within 38 metres of a waterbody or wetland area.

These twelve criteria provide a sound technical basis for the ordinance and are fundamental to its success. Another strength of the ordinance is the fact that the Town solicited, considered, and integrated comments from the development community in developing the ordinance.

From a practical standpoint, the ordinance has had a significant impact on planning of development projects, the design and review process, and construction practices in the Town. The 3-Peaks Golf Course/single family residential development project in the Town serves as a good example of application of the ordinance. The 235-hectare development consists of an 18-hole championship golf course, clubhouse, and associated course maintenance facilities and 397 residential units. There are many wetlands on the site located as pockets spread over the development parcel. Development of the parcel is further constrained by steep slopes in many areas. To-date, the developer has prepared three detailed Disturbance Permit Applications (DPAs) to demonstrate compliance of project phases with Ordinance 1999-1. These DPAs have included lot-by-lot assessments of potential adverse impacts to buffer zones and adjacent wetlands and waterbodies and have provided detailed mitigation measures and BMPs to reduce the width of the variable outer buffer. Planned individual building pads, driveways, roads, and other lot and infrastructure features have been moved or otherwise altered to protect buffer zones and/or compensate for impacted buffer zone functions. Through careful planning, the total area of wetland disturbance for the 235-hectare site is less than 0.25 hectares, which will be compensated for by mitigation at a 2:1 ratio. The developer has frequently met with Town staff in the field to discuss compatibility of the proposed development with the requirements of the ordinance.

3. Grant Ranch Stormwater Agreement

The Grant Ranch development is a large single-family residential development in the western Denver metropolitan area in Littleton, Colorado. A portion of the Grant Ranch development falls within the watershed of the Bow Mar Reservoir. The Bow Mar Reservoir is owned by the community of Bow Mar, a relatively affluent municipality of approximately 900, which utilises the reservoir for swimming, fishing, boating, and other beneficial uses. Classified uses of the reservoir include [7]:
- Aquatic Life 1, Warm
- Recreation 1
- Agriculture

The portion of the Grant Ranch development proposed within the Bow Mar Reservoir watershed included approximately 200 homes. This proposal raised concerns among the residents of Bow Mar as to the potential for adverse impacts to Bow Mar Reservoir from runoff from the residential development. These concerns were addressed in negotiations between the developer and the Bow Mar Homeowners, which culminated in a March 1997 Stormwater Agreement [5]. The Stormwater Agreement required the developer and the metropolitan district serving the Grant Ranch to maintain advanced erosion and sediment controls during construction and to implement permanent structural and non-structural BMPs following the completion of construction.

Numeric criteria were established to gauge compliance with the terms of the agreement for both construction and post-construction phases of the development. During construction, the developer was required to meet a total settleable solids discharge limit of 2.5 mg/L/hr. Post-construction discharge limitations were more extensive, addressing parameters including:

- Phosphorus - Total Phosphorus and Dissolved Orthophosphate
- Nitrogen - Total Nitrogen, Nitrate, and Nitrite
- Chemical Oxygen Demand
- Total Suspended Solids
- Fecal Coliform
- Chloride
- Total Recoverable Cadmium, Chromium VI, Copper, Lead, Manganese, and Zinc
- Oil and Grease
- Ethyl Benzene
- Toluene
- Glyphosate
- Malathion

Numeric criteria were set based on a number of factors including established State and Federal water quality standards [7,8,10], performance studies of similar treatment systems in the Denver metropolitan area [11], and values from the national literature. Where existing technical guidance was not sufficient for establishment of numeric criteria for a parameter, the agreement provided for establishment of criteria based on water quality data collected during the initial years of monitoring of the system.

The criteria for these parameters are required to be met under both dry and wet weather conditions, and compliance is assessed based on a total of eight sampling events, two wet events and two dry events during the periods of March to May and June to September, respectively. Wet weather events that are used to gauge compliance with the agreement correspond to the more frequently occurring, runoff producing storms with rainfall between 2.5 and 12.7 millimetres. Fully automated sampling stations, equipped with modems, raingauges, and automatic sample collection and flow measurement equipment, have been established at the site to collect the samples necessary to assure compliance with the agreement.

To meet the during-construction discharge limitation of 2.5 mg/L/hr for settleable solids, the developer utilised BMPs including:

- Three large sedimentation basins.
- Conventional erosion and sediment control measures including straw bales, silt fence, and inlet protection.
- Source control construction practices including "Good Housekeeping" measures.
- An innovative system by which treated runoff from the sedimentation basins was pumped to the top of a meadow and released to flow as sheet flow through the meadow. This treatment system provided treatment through infiltration of stormwater and filtration of sediments as the sheet flow passed through the meadow.

92

The performance of the treatment system during construction was excellent, with only one incident of non-compliance with the numeric limit, which corresponded to a storm event with a frequency of approximately one in twenty-five years. The construction phase of the project ended in the fall of 1999.

To meet the strict criteria for post-development conditions, the developer utilised a combination of structural and non-structural BMPs. The developer and design engineer opted to utilise BMPs in series, with initial treatment provided by three extended dry detention ponds followed by secondary treatment in a wetland/water quality pond. A plan view of this treatment system is shown in Figure 2. A design storm corresponding to the 80th percentile annual event was used to determine the WQCV. Extended dry detention basins were designed to drain the WQCV over a 40-hour period and the wetland/water quality pond had a design residence time of 24 hours, resulting in a 64-hour residence time for the system for the WQCV. The Stormwater Agreement included a commitment from the metropolitan district serving Grant Ranch and the Grant Ranch Homeowners to maintain these BMPs. Nonstructural BMPs specified in the agreement include collection and disposal of lawn clippings; restrictions on the use, storage, and disposal of herbicides and pesticides; limits on the use of sand, salt, and other de-icing agents for snow and ice control in the winter; and vegetation management practices.

Figure 2. Permanent BMP Treatment System for Grant Ranch

4. Conclusion

Local regulations and agreements for protection of water quality and control of NPS pollution are becoming increasingly common. These ordinances and agreements are, in effect, forms of non-structural BMPs designed to prevent and reduce pollution and protect water quality through creation of processes for planning and review, establishment of requirements for structural and non-structural BMPs, and provision of mechanisms for assuring compliance.

5. References

1. United States Environmental Protection Agency (USEPA) (1998) *National Water Quality Inventory: 1998 Report to Congress* (EPA 841-R-00-001), USEPA Office of Water, Washington, D.C.
2. State of Colorado (2000) *Preliminary Population Projections for Colorado Counties, 1990-2025*, Colorado Department of Local Affairs, Demography Section of Colorado Division of Local Government, http://www.dlg.oem2.state.co.us/demog/project.htm.
3. Cherry Creek Basin Water Quality Authority (1999) *Cherry Creek Reservoir Watershed Stormwater Quality Model Ordinance*, Adopted October 21, 1999.
4. Town of Silverthorne, Colorado (1999) *Ordinance 1999-1 Amending Chapter 4 of the Silverthorne Town Code to Establish Waterbody, Wetland, and Riparian Protection Regulations*, Adopted January 1999.
5. Town of Bow Mar; Bow Mar Owners, Inc.; Lower Bowles Company; The Joseph W. Bowles Reservoir Company; RSRF Ranch Company, LLC; and Bowles Metropolitan District (1997) *Stormwater Agreement Dated March 14, 1997*.
6. State of Colorado (1998) *Status of Water Quality in Colorado 1998 (305(b) Report)*, Colorado Department of Public Health and Environment, Water Quality Control Division.
7. State of Colorado (2000) *Regulation Number 38 - Classifications and Numeric Standards for South Platte River Basin; Laramie River Basin; Republican River Basin; Smoky Hill River Basin*, Originally adopted April 6, 1981, last update effective June 30, 2000, Colorado Department of Public Health and the Environment, Water Quality Control Commission.
8. State of Colorado (1997) *Reg. 31 - Basic Standards and Methodologies for Surface Water*, Originally adopted May 22, 1979, last update effective March 2, 1999, Colorado Department of Public Health and the Environment, Water Quality Control Commission.
9. Denver Urban Drainage and Flood Control District (1999) *Urban Storm Drainage Criteria Manual, Volume III*, UDFCD, Denver, Colorado.
10. United States Environmental Protection Agency (USEPA) (1998) *Ambient Water Quality Criteria for Chloride* (EPA 440/5-88-001), USEPA Office of Water, Regulations and Standards, Criteria and Standards Division, Washington, D.C.
11. Urbonas, B., Carlson, J., and Vang, B. (1993) Performance of the Shop Creek Joint Pond-Wetland System, *Flood Hazard News*, 23:1.

GIS-SUPPORTED STORMWATER SOURCE CONTROL IMPLEMENTATION AND URBAN FLOOD RISK MITIGATION

C. MAKROPOULOS, D. BUTLER & C. MAKSIMOVIC
Imperial College of Science, Technology & Medicine,
Department of Civil & Environmental Engineering,
London, SW7 2BU, UK.

1. Introduction

The natural response of a catchment, characterised by infiltration, evaporation, attenuation, surface storage and reduced runoff [1], is altered by the urbanisation process. The construction of buildings, roads, and pavements increases the impermeability of the catchment and thus the volume of water leaked into the subsoil is significantly reduced. Furthermore, the changes in land use involve loss of vegetation, which consequently blocks the evapotranspiration mechanism of stormwater reduction. The classic "end-of-pipe" approach to these problems, which is characterised by improvement in the capacities of streams and drainage ditches, actually increases the velocity of flow. This reduction of the catchment response time consequently increases the maximum rate of flow discharging to the drainage system and finally increases the frequency of significant floods [2]. An alternative to this approach is source control promoting stormwater reuse and application (waste minimisation, reuse, recycle). Researchers dealing with stormwater drainage and flood risk management have been increasingly interested in source control [1, 3, 4], being in accordance with the basic concepts of sustainable drainage as described by Butler and Parkinson [5], aiming at minimising the amount of runoff by infiltration for aquifer recharge, utilising the natural (or already existing) pathways of the catchment for stormwater drainage and reducing risk of flooding using non structural means where possible as the latter have proved to be at best partial solutions. This can be achieved by a reduction in rainfall runoff volume (infiltration techniques) as well as by an increase of time to peak (grass swales, which slow down runoff).

In the multi-criteria problem of source control application, the need for efficient tools, which can facilitate the process of taking into account an ever-increasing amount of information, understanding the processes involved and quantifying the results of the analysis, is crucial. Geographical Information Systems (GIS) and Advanced Simulation Models can be of great help in this aspect, not only with their ability to handle a large amount of information for a given area, but also by predicting the possible effect of various mitigation schemes as well as presenting the results in a readable form. This in turn allows the establishment of a "common language of communication" between municipal authorities, professionals and water

J. Marsalek et al. (eds.), Advances in Urban Stormwater and Agricultural Runoff Source Controls, 95–105.
© 2001 *Kluwer Academic Publishers. Printed in the Netherlands.*

user associations [6]. The fact that experience shows that there is no general way of implementing such solutions and therefore **a site-specific implementation** is needed, makes the role of GIS and Simulation Models all the more crucial. Although source control application can have a significant impact on flood risk, the effect is neither easily quantifiable, nor easily observed in its true, spatially distributed nature. The actual problem Planners and Local Authorities have to solve is be able to quantify the effect of a site specific implementation of a source control scheme with its spatial variability and identify improvement or deterioration in each part of the application area. This study illustrates a method to solve this problem, using GIS (IDRISI-GIS) to identify application areas, a drainage model (MOUSE 4.0) to predict changes of flow in the drainage system and GIS again as a result presentation tool.

2. IDRISI – GIS

IDRISI version 2 for Windows, which was used as the principal GIS software in this study, is a user-friendly, easy to use, raster-based system, which provides the user with a variety of tools for presentation, data analysis and decision support. IDRISI includes a number of built-in functions for the manipulation of spatial information, a detailed description of which can be found in IDRISI User's Guide [7]. A number of these functions (modules) was utilized in this study, ranging from simple ones like ASSIGN, EXTRACT, RECLASS, to more complex ones like the Image Calculator and the FUZZY and Multi-Criteria Evaluation (MCE) modules.

3. MOUSE Drainage Simulation Model

MOUSE divides the drainage system in two parts: the catchment and the network [8, 9, 10]. As input, MOUSE requires data of the physical properties of the system and of the boundary conditions. The modelled area is divided into a number of sub-catchments and each sub-catchment is attached to a node of the network. The physical and hydrological properties of each sub-catchment are entered according to the particular runoff model used. Two surface models are available for runoff simulation: the time area method and the kinematic wave method. The kinematic wave method, which was used for this study, uses Manning's equation and calculates losses due to evaporation, wetting, storage, and infiltration (Horton's equation). The pipe flow model can simulate unsteady flow with both free surface and pressurized flow conditions. It is based on a finite difference numerical solution of the one dimensional, free surface flow equations (Saint-Venant). Vertically homogeneous flow conditions are assumed and the network is solved for open channels as well as pipes. Flow features such as backwater effects and surcharges can also be simulated. Pressurized pipes are calculated as free surface using a narrow (Preissman) slot as a vertical extension to the pipe cross section. It is to be noted that MOUSE 4.0 has a serious disadvantage when dealing with open channels instead of buried pipes. The simulation part as well as the tools developed for buried pipes (i.e. flood computation) cannot really cope with the (more complex) problems of open pipes and a number of interventions from the user have to be made in order to arrive at acceptable results (i.e. raising the height of the channels that flood, to take into account the additional cross-section of the street). In addition, source control application is not incorporated in this version of the model and must be applied by changing the pervious/impervious ratio and the infiltration properties of the catchment, which is a simplification of the natural phenomenon.

4. Description of the Case Study Area

Novi Sad is a medium sized city to the north of Belgrade and is characterized, as many other central European cities by non-uniform urban development, evidence of different phases of construction. The river Danube separates the old city from the new city and on the outskirts of the new city exists a large peri-urban area, characterized by uncontrolled urbanization and industrial activities. The catchment has one of the oldest drainage networks in the Balkans. Most of the area underwent uncontrolled urbanization without proper infrastructure development, which resulted in increased runoff, flooding and inconvenience to the inhabitants. This has resulted in calls for rehabilitation of the network, with the public pressing for installation of buried pipes as this option is considered more modern and efficient. However, the buried pipes option is both expensive and increases peak flows and decreases time to peak, thus overloading the local receiving water bodies. The current study, as well as previous ones [11], shows that the sustainable drainage option would be to **rehabilitate the existing drainage ditches** by imposing **detention** in the channels, increasing **infiltration (source control)** and enhancing **stormwater quality** with the use of constructed wetlands. Due to insufficiency of actual data, information that was not known for the case study catchment (groundwater level, soil characteristics, connectivity and income) was hypothesized and used for further assessment.

5. Application

5.1. PRE-PROCESSING

Field data were introduced into the geographical database; these include groundwater depth, topsoil type, slope, vegetation, land use, distance from outlet and inhabitants income. Fuzzy logic provided a standardisation procedure through the application of appropriate membership functions. The standardised output of the initial step in GIS layers was introduced in the IDRISI multi-criteria analysis module (MCA) and the output suitability map was obtained using an ordered weighted average technique (OWA) [7, 12]. Mapping of source control suitability is discussed in more detail elsewhere [13]. A specific set of source control methods, known as "infiltration techniques", including infiltration, percolation beds and grass swales [14], was selected, as an example, to further investigate the impact of the proposed source control application on the catchment by means of simulation.

98

To identify the exact sub-catchments where source control (in this case infiltration techniques) would be more applicable, the infiltration technique suitability map (Figure 1) was used together with the sub-catchments layer. The EXTRACT module was applied to compute a mean suitability value for each sub-catchment[1]. The outcome was a value file that was then ASSIGNED to the sub-catchment layer. The new layer was thresholded (accepted areas for application had suitability ratings over 230/255) and the RECLASS function was utilized to create a Boolean image identifying the application zone. The result obtained can be seen in Figure 2. The area for the application was calculated to be 159.2 ha (total catchment area 644.1 ha) or about 25% of the catchment. The Boolean image containing the application map was then converted from GIS image format to bitmap format and the resulting file imported to MOUSE. It was used as a background image to the network using the correct co-ordinates and the sub-catchments in question were selected using the polygon selection facility provided in MOUSE 4.0.

Figure 1. Infiltration techniques suitability map and delineated sub-catchments.

Figure 2. Final area of infiltration techniques application (shaded), over the digital elevation model.

---

[1] The suitability maps were derived as a continuum, thus each pixel had its own suitability value. To be able to select one value per sub-catchment, a summary statistic (mean, median etc) had to be used. IDRISI GIS enables such an aggregation analysis to be performed.

5.2. INPUT TO MOUSE

The next step was to alter the sub-catchment characteristics to take into account source control application and in particular infiltration techniques. At this stage, four possible methods were identified (taking into account the capabilities of the model).

1. Change the impervious areas of the selected sub-catchments to pervious[2].
2. Subtract a pre-calculated amount of rainfall (to take infiltration and surface storage into account) and then apply the rest on the selected sub-catchments[3], while applying all of the rainfall to the rest of the catchment.
3. Exclude the selected catchments from the simulation altogether[4].
4. Increase the value of infiltration parameters (i.e. Horton's f), to take into account the increase in infiltration capabilities of a soil when structures such as infiltration blankets etc. are used.

For the purposes of this particular study, it was decided that the change of the impervious areas of the selected catchments to pervious would take into account the effect of infiltration techniques (i.e. disconnection of roofs and redirection of runoff from impervious areas onto pervious). The fact that, in reality, the total pervious area remains the same, i.e. rainfall should be applied to the whole sub-catchment while infiltration should occur only to the originally pervious part, can to some extent be balanced by the fact that infiltration techniques would actually improve the infiltration capabilities of the soil, a factor that is not taken into account here. The fact remains that the results shown here should be viewed as somewhat optimistic and a modelling approach better incorporating source control would certainly be of interest for future research.

5.3. THE RAINFALL DATABASE

The rainfall database was created based on statistical analysis of a historical rainfall for the specific catchment. The charts were obtained from previous research [11] on the case-study area, and can be seen in Figures 3 and 4.

---

[2] The physical meaning of such an approach can, for instance, be the disconnection of roof downpipes, allowing the water to pass over a pervious area and be partially infiltrated. Even though the roof is still impervious, as far as runoff is concerned it then behaves as pervious.

[3] The sub-catchments where source control is to be implemented

[4] The physical meaning of such an approach is that all rainfall is considered to have infiltrated. This presupposes that there is enough storage capacity in the infiltration or detention structures not to allow any runoff.

100

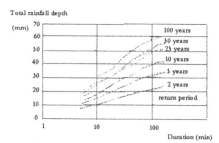
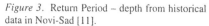

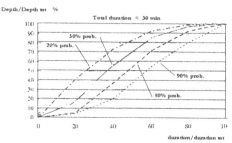

Figure 3. Return Period – depth from historical data in Novi-Sad [11].

Figure 4. Hyetograph with a given shape probability for Novi-Sad [11].

Two rainfall events were applied to the catchment with and without infiltration techniques, thus creating four groups of results. The first rainfall had a 2-year return period and the second 10 years. In both cases, the duration was, as an example, 30 min and the shape hyetograph probability was 90%.

6. Network

The catchment itself consists of 953 sub-catchments and an equal number of nodes plus two outlets, which are linked together with 957 conduits. The majority of the conduits are open channels, but there are also a number of buried pipes wherever the channels are crossed over by the traffic network. The channels are trapezoidal with dimensions[5] ranging from 10 cm X 10 cm to 150 cm X 150 cm. Diameters for the pipes range from 300 mm to 1000 mm. The large variety of materials and types of channels used, the design of the network subject to changes and additions over the years, as well as the changes in slopes, result in complex hydraulic phenomena (backwater effects and differences in patterns between maximum water levels, times to peak and discharges). The analysis of these phenomena – although interesting from an engineering point of view – is beyond the scope of the present study but has been studied in detail in [11] without source control simulation. There were, however, hydraulic phenomena that had to be dealt with in the course of the simulation: MOUSE 4.0 (and all the previous versions) cannot cope with overflowing channels and negative slopes. In order to overcome this difficulty the original network was slightly modified in two ways.

The sidewalls of channels that overflowed (typically those upstream with dimensions of 10 X 10 or 30 X 30 cm) were artificially raised to the necessary height. This takes into account the fact that the cross-section has been reduced over the years from its original size due to soil deposits. This also takes into account the additional cross section (the road) used if the channel floods. Even if the sidewalls of the channel are overtopped this doesn't automatically imply that a flood occurs. Only if ground level is reached will the flood occur. It is thus difficult to identify the areas that are

---

[5] Their original size was significantly larger but the channels were silted due to lack of proper maintenance.

flooded after the simulation ends, because version 4.0 of the model does not provide such a function.

To obtain simpler behaviour of the system from a hydraulics point of view, an outlet was added to the system where an older outlet (now out of use) was previously situated. The reasons leading to the closing of the outlet in practice were relevant to urban planning decisions in the area and had nothing to do with the methodology for source control application created here.

7. Simulation

Each rainfall event selected for simulation was applied to the catchment with and without source control application resulting in four sets of data. Two simulation models were used to derive each set. The runoff model (kinematics flow approximation) was initially used to produce hydrographs, which in turn acted as input to the nodes for the network flow model. The simulation time for the runoff model was 1 h and for the pipe flow model 2 h. MOUSE 4.0, with the help of two additional programs, MOUSE View and MOUSE Print, facilitates the presentation of results of a single simulation. In order to visualise the spatial change in maximum water levels and discharge after the application of source control, the output water levels in channels and discharges were imported to EXCEL. An indicator of change was calculated for maximum water level (Hmax) in the particular channel:

$$\frac{H \max{}_{no\_source\_control} - H \max{}_{source\_control}}{H \max{}_{no\_source\_control}} \tag{1}$$

and another for maximum discharge (Qmax):

$$\frac{Q \max{}_{no\_source\_control} - Q \max{}_{source\_control}}{Q \max{}_{no\_source\_control}} \tag{2}$$

The results were then exported to IDRISI as value files and the INTERPOL module used to calculate a surface indicating the reduction in the risk of flooding in the area due to source control application. The results of the simulation can be seen in Figures 7-10 indicating reduction in maximum discharge and maximum water level for rainfall events with 2 and 10 year return periods.

Figure 11 shows a typical hydrograph of runoff from the catchment to the drainage network, both before and after implementation of source control.

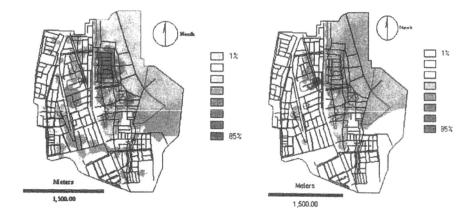

Figure 7. Reduction in maximum water level for 2 year rainfall.

Figure 8. Reduction in maximum water level for 10 year rainfall.

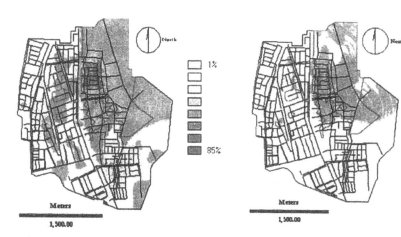

Figure 9. Reduction in maximum water discharge for 2 year rainfall.

Figure 10. Reduction in maximum water discharge for 10 year rainfall.

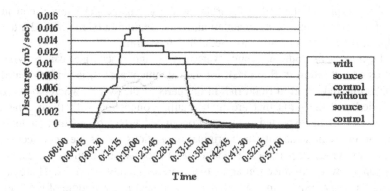

Figure 11. Typical hydrographs pre- and post-source control implementation.

8. Discussion of Results

The results on inflow volume from the catchment to the drainage system indicate a significant reduction due to application of infiltration techniques. In the case of the 10 year storm, the reduction in volume is of the order of 7% while in the case of the 2 year storm the reduction is of the order of 12.5%. To the extent that the model is accurate, the results clearly indicate that the effect of infiltration techniques is more significant in rainfalls of short return period. This was to be expected as infiltration structures (trenches etc.) have a given volume of storage and a given rate of infiltration that can be attained. Infiltration structures are normally designed for a specific return period and part of the rainfall will become runoff if the return period is exceeded or if the rainfall is repeated and the structure is already part full. A balance between the cost of constructing more storage capacity for the infiltration structures and constructing a larger drainage network (with all its implications) has to be achieved, wherever such techniques are to be implemented [15, 16, 17].

The fact that source control seems more effective in rainfalls of shorter return periods is all the more evident in the IDRISI maps (Figures 7-10). The upper part of the catchment experiences the most significant reduction both in H max and Q max. The net effect ranges between a maximum of 40% to 76% in reduction of water levels for 10 and 2 year return periods respectively and 60% to 80% in reduction of discharges for 10 and 2 year return periods. Because only one method of source control was simulated in this study, it would be particularly interesting to see the results from the simulation of application of other methods in different parts of the catchment (in accordance with the relevant suitability maps) that have already been calculated during this research), which should result in more pronounced reductions in both outlets and particularly the lower one (i.e. application of porous pavements in central areas).

104

The graphs of the inflow hydrographs (e.g. Figure 11) show a considerable delay, after source control implementation, in the beginning of runoff and attenuation in the peak flow rate, in addition to a reduction in volume.

Decreases in both maximum water levels and discharges can be viewed as indicators of **reduction in flood risk** for the catchment area. It is evident that there is a considerable reduction in flood risk in the parts of the system where source control was applied and the parts that are in hydraulic connection with them. The results show, as was to be expected, a more pronounced effect of infiltration techniques application in urban areas with high original imperviousness as the net effect from the change (impervious to pervious) is much more pronounced than in the cases where source control was introduced in areas that were originally pervious. There seems to be a slight negative impact in a specific sub-catchment in the middle of the study area. It is not possible to explain this without detailed hydraulic analysis of the specific part of the network. It can be considered however as an irregularity of the network at this particular point rather than an effect of source control application. This further illustrates the point that the GIS-based methodology developed in this study takes into account the specific characteristics of the target area, thus serving as a support tool for site-specific source control implementation, as well as an assessment of effect that can be understood by specialists and non-specialists alike. The fact that a number of people that would be involved in a source control implementation and flood risk reduction scheme would be non-specialists (i.e. Local Authority workers) makes this aspect all the more significant.

The site-specific source control planning process illustrated above can be used further to take into account quality issues recently associated with source control, making efficient use of additional data such as traffic load, roof material etc. to support the choice of the proper source control strategy for each part of the system [18]. This would help to ensure that infiltration and storage is optimum both from a quality and a quantity perspective.

9. References

1. Andoh, R.Y.G. (1994) Urban drainage - the alternative approach. *Proceedings of 20th WEDC Conference*, Sri Lanka, pp. 147-150.
2. Smisson, M. (1979) A review of the stormwater drainage of new developments, *IPHE*. **8 (1)**, 64 – 73.
3. Ellis, J.B. (1995) Integrated approaches for achieving sustainable development of urban storm drainage, *Wat.Sci.Tech*. **32 (1)**, 1-6.
4. Kollatsch, D. (1993) Futuristic ideas to create a most efficient drainage system, *Proceedings of 6th Int. Conf. On Urban Storm Drainage*, Niagara Falls, Ontario, pp. 1231-1236.
5. Butler, D. and Parkinson, J. (1997) Towards sustainable urban drainage, *Wat.Sci.Tech*. **35 (9)**, 53-63.
6. Makropoulos, C. (1997) Sustainable cities: GIS support to the analysis of source control application in urban storm drainage networks, MSc Dissertation, Imperial College, London, 121 p.
7. Eastman, J.R. (1997) User's guide, IDRISI version 2 for Windows, Clark University, Worchester.
8. DHI, MOUSE- (1993) User Manual and Reference Manual, Danish Hydraulic Institute, Hörsholm, Denmark.
9. DHI, MOUSE- (1994) User Manual and Reference Manual, Danish Hydraulic Institute, Hörsholm, Denmark.
10. DHI, MOUSE- (1996) User Manual and Reference Manual, Danish Hydraulic Institute, Hörsholm, Denmark.

11. Milutinovic, R. (1996) Project of Reconstruction of the Klisa Catchment, Hydrobureau and Institute of Hydraulic Engineering, Civil Eng. Dept., University of Belgrade, Belgrade.
12. Maltczewski, J. (1999) *GIS and Multicriteria Decision Analysis,* John Wiley & Sons Publishers, New York.
13. Makropoulos, C., Butler, D., and Maksimovic, C. (1999) GIS supported evaluation of source control applicability in urban areas, *Wat.Sci.Tech.* **39 (9),** 243-252.
14. Urbonas, B. and Stahre, P. (1993) *Stormwater: Best Management Practices and Detention.* Prentice – Hall Publishers.
15. Leonard, O.J. and Sherriff J.D.F. (1992) Scope for control of urban runoff, CIRIA, Report 124, London.
16. Roberts, J.D. (1995) Infiltration Drainage – appraisal of costs, HR Wallingford and Ove Arup Partners, CIRIA, Report 24, London.
17. Declerck, C. and Andoh, R.Y.G. (1996) The cost effective approach to stormwater management? Source control and distributed storage. *Proceedings of 7th Int. Conf. On Urban Storm Drainage,* Hannover, pp. 1743-1749.
18. Pitt, R., Shirley, C., and Field, R. (1999) Groundwater contamination potential from stormwater infiltration practices, *Urban Water* **1 (3),** 217-236.

PROTOCOLS AND METHODS FOR EVALUATING THE PERFORMANCE OF STORMWATER SOURCE CONTROLS

M.M. QUIGLEY[1] & E.W. STRECKER[2]
GeoSyntec Consultants
[1]*532 Great Road, Acton, MA 01720, U.S.A.*
[2]*333 SW Fifth Avenue, Suite 600, Portland, OR 97204, U.S.A.*

1. Introduction

Logic suggests that source controls, (defined in this paper as any practice that prevents the introduction of pollutants into stormwater runoff), can, in many cases, achieve significant improvements in water quality and be more cost-effective than downstream treatment of runoff. For example, the detection and elimination of illicit connections to a storm drainage system would be expected to result in reduced pollutant discharges and would likely be significantly more effective and less costly than downstream treatment.

Assessments of programs and practices that target prevention of the introduction of pollutants to stormwater runoff are of increasingly utility to stormwater system managers. Valid and uniform measures of the effectiveness of source control measures can provide a key tool for documenting and reporting that stormwater programs are achieving or not achieving performance metrics and perhaps provide a basis for selecting, configuring, and prioritising amongst the many options for source controls as well as treatment controls. These measures can convey useful information to institutional managers, regulators, and interest groups. This paper examines approaches that have been used by investigators primarily for the assessment of the effectiveness of treatment based controls from a water quality perspective, and recommends approaches for modifying and utilising these approaches for source controls, while examining the differences that may limit their utility.

2. Goal Attainment and the Definition of Effectiveness

There are a number of goals that are typically targeted by source control assessment efforts. The typical major goals of source control programs, along with recommendations for methods applicable for evaluation of source control effectiveness, are provided in Table 1. As Table 1 indicates, the attainment of some of these goals could potentially be measured directly through monitoring of downstream water quality (i.e., outfall or receiving water quality) or biological and physical habitat. They can also be assessed through indirect means such as monitoring upstream

J. Marsalek et al. (eds.), Advances in Urban Stormwater and Agricultural Runoff Source Controls, 107–117.
© 2001 *Kluwer Academic Publishers. Printed in the Netherlands.*

water quality (e.g., water quality at the source), total prevented pollution at the source (i.e., quantity prevented from not entering runoff), or other surrogate measures of program effectiveness (e.g., number of illicit connections removed, public surveys/user perception, or independent observation of the number of instances that a particular practice was followed). Alternately, some goals can be assessed through the reporting of programmatic measures of effectiveness such as the number of meetings held or expenditures on control practices.

An example best illustrates the difference between direct, surrogate, and programmatic measures of effectiveness. The vacuum removal of standing puddles in impervious source areas during inter-storm periods has been considered as a source control measure by some municipalities for control of human pathogen indicating bacteria (coliforms) [1]. A variety of measures are available for determining the effectiveness of this type of practice. If the goal of an effectiveness assessment is to document that the practice removes meaningful quantities of a particular pollutant from wet weather discharges, monitoring of water quality of wet weather runoff may be a potential approach. However, given all other watershed sources, other BMPs, and downstream processes that may influence downstream water quality, it may be difficult to collect enough samples to statistically observe that a change had occurred due to this one practice alone. Depending on the requirements of the program, documentation of the number of puddles vacuumed or the volume of water removed may provide a valid metric for the evaluation of the potential program along with water quality analyses of samples of the water removed. Using this alternate approach, a surrogate parameter (number of puddles or volume of water removed) can provide a cost effective means for monitoring the source control practice. Estimates of the ability of a practice to remove a particular pollutant could potentially be modelled from surrogate measures of effectiveness, particularly where a model can be calibrated using field-collected samples (e.g., a pilot scale study on a small catchment).

Programmatic measures of effectiveness can be useful, particularly where multiple goals, in addition to water quality improvements, are targeted. In many cases, a program can document attainment of goals (e.g., building community awareness) through programmatic measures of effectiveness such as numbers of meetings held or pamphlets distributed. However, using programmatic measures typically does not provide a means of assessing the degree to which water quality goals for a program have been met unless a pilot scale study has been conducted that is statistically valid for modelling the relationship between programmatic effectiveness and changes in runoff water quality or aquatic biological health.

TABLE 1. Goals of source control programs and practices and the ability of specific types of effectiveness studies to provide useful information on assessment of how these goals are met.

Category of Goal	Goals of Source Control Programs and Practices	Ability to Evaluate Performance and Effectiveness
Toxicity	• Reduce acute toxicity of runoff	D, S
	• Reduce chronic toxicity of runoff	D, S
Regulatory	• Meet local, state, or federal requirements for BMP implementation	D, S, P
	• Compliance with NPDES permit	D, S, P
	• Meet local, state, or federal water quality criteria	D, S
Implementation Feasibility	• Source control's ability to function within management and oversight structure	P
Cost	• Capital, operation, and maintenance costs	P
Aesthetic	• Improve appearance of site	S, P
Maintenance	• Operate within maintenance, and repair schedule and requirements	S, P
	• Ability of control measure to be modified or expanded	S, P
Longevity	• Long term effectiveness, understand temporal changes in effectiveness	D, S, P
Resources	• Improve downstream aquatic environment/erosion control	D, S
	• Improve wildlife habitat	S
Safety, Risk and Liability	• Implemented without significant risk or liability	P
Public Perception	• Information is available to clarify public understanding of runoff quality, quantity and impacts on receiving waters	D, S, P

D - can be evaluated through direct monitoring of water quality and flow or biological and habitat assessments
S - can be evaluated through surrogate measures of source control implementation and/or effectiveness
P - can be evaluated through programmatic measures of source control implementation and/or effectiveness (e.g., costs for implementation, number of times practice was conducted)

3. Approaches for Quantitative Assessment

Quantitative assessments of source control effectiveness are useful as direct measures of effectiveness. Two main approaches, that are applicable to quantitative assessment of source controls, have been used by researchers studying the effectiveness of primarily treatment based best management practices: reference watershed studies, also known as paired watershed studies, and temporal variation (before and after) studies.

Each of these methods has advantages and disadvantages resulting from tradeoffs between control of static (factors that remain constant with time) and state variables (those that change with time). Reference watersheds allow for the same conditions for state variables (e.g., weather conditions) during a given monitoring period and require less time than temporal variation studies. However, reference watershed studies require careful site selections to ensure that the two watersheds observed have similar characteristics (e.g., watershed and source area size, similar quantity and type of sources, as well as similar drainage systems and processes). Where reference watersheds can be well matched and water quality can be measured

110

directly, paired statistical tests can be used in place of non-paired tests to demonstrate differences in the observed water quality between catchments, thus increasing the power of the statistical methods used. Recommended tests for analysis of paired data are provided in Table 2. Further information on paired watershed studies can be found in Clausen and Spooner [2].

TABLE 2. Statistical methods to use with paired watershed studies.

Test Name	Type	Benefits	Drawbacks/Assumptions
Signed-Rank Test	Non-parametric	• Does not require underlying distributional assumptions • Is tolerant of small numbers of non-detects	• Has less power than methods that assume some distributional characteristics
Wilcoxon Signed Rank Test	Non-parametric	• Does not require assumption of normality • Has greater power than Sign-Rank Test	• Is not tolerant of non-detects • Requires the underlying distribution to be symmetric • Has less power than parametric tests
Paired t Test	Parametric	• Where assumptions are valid, provides greater power than non-parametric tests	• Assumes differences follow a normal distribution

Temporal variation studies follow a study design that helps ensure that static variables (e.g., watershed size, number of households, miles of road, structure of drainage system) are similar between periods with or without source control practices in place. This study design, however, does not fully account for differences in state variables (e.g. weather and/or pollution source patterns). It is also important to determine if the static variables are well addressed by the study design. For example, if the watershed has development or re-development occurring and/or the implementation of other BMPs, than these variables may not be static. In addition, temporal variation studies require longer monitoring periods. Recommended statistical methods for comparisons of measures of effectiveness in temporal variation studies are provided in Table 3.

Ideally, the two approaches to study design can be used in concert. If apparently well-paired catchments can be found, it is best to establish that the watersheds are in fact useful for a paired watershed approach by monitoring the measure of effectiveness in both catchments prior to instituting the source control program or practice. This period must be long enough to establish statistically that the observed measure of effectiveness is similar or that the differences in static variables can be taken into account prior to source control implementation. For some types of source controls, the control and test watershed can be alternated to improve the robustness of the study results.

TABLE 3. Statistical methods to use with temporal variation studies.

Test Name	Type	Benefits	Drawbacks/Assumptions
Rank-Sum Test (Wilcoxon ran-sum test or Mann-Whitney test)	Non-parametric	• Does not require underlying distributional assumptions • Results of test are not affected by transforming data	• Has less power than methods that assume some distributional characteristics
Two-Sample *t* Test	Parametric	• Where assumptions are valid, provides greater power than Rank-Sum Test	• Less powerful where data are skewed or non-normal. Rank-Sum may be more powerful in these situations.
Analysis of Variance (ANOVA)	Parametric	• Can be used where more than two data sets are available (extension of the t Test) • Can test for the effect of more than one factor	• Less powerful where data are skewed or non-normal. Rank-Sum may be more powerful in these situations.

4. Detecting Water Quality Changes Resulting from Source Control Implementation

In many cases a particular source control measure assessed individually may have a small impact on water quality. The smaller the impact, the more difficult it becomes to observe the effect of the program or practice individually. This is due to statistical constraints on the number of sample observations required to estimate central tendency and variability in a population. If the goal of an effectiveness assessment program is to determine if a single control practice significantly decreases the load or concentration of a constituent reaching a receiving water during wet weather flows and it has been determined that water quality sampling will be required to achieve project goals, then enough samples must be collected to support a hypothesis that the measure of effectiveness (e.g., mean or median event mean concentration) from a catchment where the source control has been implemented is significantly different from a catchment where the source control is absent. Given that the measure of effectiveness can be assumed to be normally distributed and that the coefficient of variation (standard deviation/mean) for the parameter observed is similar for both the treatment and the non-treatment catchments, a nomograph can be generated to provide an estimate of the number of sample observations required to observe a given percent difference between the mean values of the two samples (see Figure 1). A similar approach can be used for log-normal data [3]. In general, equation 1 can be used to determine the number of samples required for situations where the coefficient of variation of the measure of effectiveness is not equal at the treatment and the non-treatment catchments.

$$n = \left[Z_{\alpha/2} \frac{(COV_1 + COV_2 - COV_2 \times \%difference)}{\%difference} \right]^2 \qquad (1)$$

where,

n: number of samples required

Z: area under normal distribution

α: significance level (confidence level equals 100*(1 - α)%); α = 0.05 indicates
        a 95 percent confidence level

COV_1: coefficient of variation of measure of effectiveness for catchment 1 or period 1

COV_2: coefficient of variation of measure of effectiveness for catchment 1 or period 2

This equation was derived by 1)setting the lower limit of the 95% confidence interval of the observed measure of effectiveness in the non-treatment catchment equal to the upper limit of the 95 % confidence interval of the observed measure of effectiveness for the treatment catchment (i.e., level of significance of 95% and power of 95%); 2)substituting the coefficient of variation multiplied by the mean for the standard deviation; 3)further substituting the difference between the means, for the mean, at the treatment catchment; and 4)rearranging the equation to solve for the number of samples.

Estimates for the coefficient of variation for water quality parameters can be obtained from a variety of sources including other monitoring work conducted on-site or at similar locations or from studies conducted by others [4,5].

Given the constraints on the ability to measure the effects of source controls due to the often small improvements resulting from any single practice, it is recommended that quantitative assessments of effectiveness, particularly pilot studies, should attempt to combine a number of practices that together have the potential to dramatically reduce pollutants introduced into runoff. The better a source control program that includes a number of controls, functions to prevent pollutants from entering runoff, the more likely an assessment program will be able to identify or provide a statistically valid estimate of measurable changes in runoff water quality. In many cases, the costs of determining if some individual practices are effective are such that if in reality only 4 of 6 practices employed were truly effective, it may cost significantly more to determine this than to just utilise all 6 measures and consider them effective as a program.

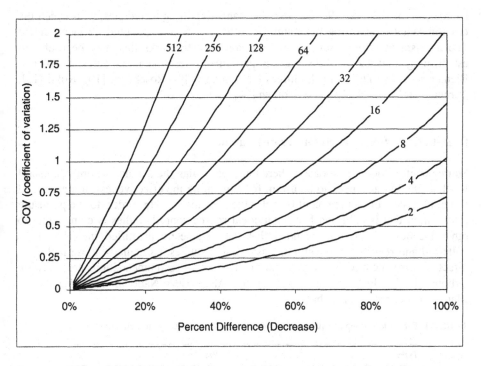

Figure 1.   Required number of observations for detecting a statistically significant percent difference as a function of the coefficient of variation (95% confidence).

5. Analysing and Reporting Indirect Measures of Source Control Effectiveness

Indirect or surrogate measures of source control effectiveness can be used effectively to document downstream runoff or receiving water quality improvements. This approach trades cost savings from not collecting large numbers of field samples for increased expenditure on verifying relationships between surrogate parameters and downstream water quality. Typically, relationships between surrogate measures of water quality and actual changes in water quality are estimated from models of the physical relationship between the two or extrapolated from pilot scale studies. In either case, valid measures of the uncertainly in each variable used in either a model or an extrapolation procedure should be documented and a sensitivity analysis should be conducted. This approach provides the only quantitative means of comparing the results of one such study to another of similar type on a uniform basis.

Where a source control assessment study is limited to collecting surrogate measures of effectiveness that may not be able to be related to water quality, for example education programs, it is highly recommended that prior to implementation of the source control and assessment program, (e.g., for example a survey of users), technical and public relations personnel work closely together to ensure that the measure of effectiveness will be a statistically valid assessment of the effects of the source control program. Where possible, standard polling techniques, with which the environmental engineering community is often not familiar, should be incorporated into assessment efforts. Several authors have written works that may be useful in establishing statistically valid surveys of participants, resource users, and stakeholders: Biemer *et al.* [6]; Dillman [7]; Groves [8]; Lockhart [9]; Rosenberg [10]; Rossi [11]; Turner [12]; Warwick [13]; and Sudman [14].

6. Data Reporting to Facilitate Transferability

Source control monitoring data gathered at a particular site should not only be useful for that site, but also need to be useful for comparing the effectiveness of similar and different types of source controls at other locations. In order to help ensure information that is collected from individual source control studies is comparable to other locations, it is recommended that the parameters identified in Table 4 be collected and reported in quantitative assessment studies, particularly those including direct or surrogate monitoring of water quality. These recommendations are based on Urbonas [15], which are the basis for the ASCE/EPA National Stormwater Best Management Practices Database.

TABLE 4. Parameters to report with quantitative assessments of source control effectiveness.

Parameter Type	Parameter
Tributary Watershed	Tributary watershed area, average slope, average runoff coefficient, length, soil types, vegetation types
	Total tributary watershed impervious percentage and percent hydraulically connected
	Details about gutter, sewer, swale, ditches, parking, roads in watershed
	Land use types (res., comm., ind., open, etc.) and acreage
General Hydrology	Date and start/stop times for storms occurring during monitoring period
	Runoff volumes for storms occurring during monitored period
	Peak 1-hr intensity for storms during monitoring period
	Average annual values for number of storms, precipitation, snowfall, min./max. temp., from appropriate weather station(s)
Water Quality	Measured or modelled event mean concentrations for targeted pollutants (see Strecker [16])
	Alkalinity, hardness and pH for each monitored storm, as appropriate
	Sediment settling velocity distribution, when available, and as appropriate
General	Type and frequency of maintenance in watershed
	Types and locations of monitoring instruments
Other Treatment	Relationship to other BMPs (structural and non-structural, and other source controls)

Based on Urbonas [15].

Water quality information that is collected directly can be collected in a standard format. The protocols established by the ASCE/EPA database fulfil this need well. New protocols need to be established for the storage of modelling results that include information about the inherent uncertainty in estimated values. This might include explicit documentation of the probable range of model results stemming from a sensitivity analysis.

Standard data reporting protocols for specific source control program types need to be established. In developing the ASCE/EPA National Stormwater Best Management Practices Database, it was found that for each practice type a separate standard data reporting format needed to be developed to accurately document the information that was most applicable to assessing that practice type. Potential protocols for street sweeping and public education are provided in Table 5.

TABLE 5. Potential information reporting protocols resulting from street sweeping and education of landscape contractors.

Source Control	Information to Report
Street Sweeping	Street surface area swept, lane miles, length of curb
	Total street surface area, total lane miles, total length of curb, and schedule/frequency of sweeping
	Amount (volume and mass) and particle size fractions of material and litter removed
	Pollutants associated with size fraction by weight
	Street surface types (AC, PCC, etc.), extent of cracking, age
	Inlet types and configuration (spacing, deep sump, standard design)
	Type/model/action of sweeper(s)
	Amount (volume and mass) of material left behind at curb, particle size
	Flushing utilised in tandem with sweeping
Education of Landscape Contractors	Landscaping practices addressed
	Survey of contractors prior to and after training regarding practices
	Estimate of total number of contractors and contractors trained
	Survey of amount of fertilisers, pesticides, and herbicides utilised in area before and after training
	Estimate of amount of areas re-landscaped or landscaped per unit area
	Estimate of the area that is maintained by professional landscaping contractors versus that maintained by individuals
Generic Parameters for All Practices	Types of activities and sources addressed
	Extent of these activities and sources (areas)
	Evaluation of pre- and post changes in activities and sources
	Number and frequency of applied measure (e.g., number of people trained and amount of time spent in training and re-training)

7. Summary and Recommendations

There is a great need to have consistency with the constituents and methods utilised for assessing source control effectiveness. It is recommended that researchers who undertake source control effectiveness studies consider the recommendations suggested here and other recommendations developed based upon further analysis of this subject.

Regulatory programs such as the United States Department of Environmental Protection's National Pollutant Discharge Elimination System have been confronted with the prospect of issuing and overseeing the implementation of permits for large numbers of separate storm sewer systems where source controls will be one of the primary management practices implemented. Without mandated water quality monitoring requirements (the Phase II NPDES program does not require water quality monitoring) and specified and consistent "measurable goals" for the evaluation of compliance with permit requirements, it will be difficult to assess source control effects on receiving water quality and for stormwater managers to make fully informed decisions about the costs and benefits of source control program implementation. It is the authors' opinion that local, state and federal agencies should proactively move toward improving the number and quality of source control assessments by funding initiatives to examine the effects of these practices, which utilise standard methods similar to those suggested here, together with guidance on study design, data collection, analysis, and methods to improve data transferability.

8. References

1. Forest, C. (2000) URS Corporation, personal correspondence.
2. Clausen, J.C. and Spooner, J. (1993) Paired watershed study design, Environmental Protection Agency, Office of Wetlands, Oceans, and Watersheds. EPA/841/F-93/009.
3. Hesse, E.T. and Quigley, M.M. (2000) Internal GeoSyntec Consultants memorandum for development of protocols for the ASCE/EPA BMP Monitoring Guidance Manual.
4. U.S. Environmental Protection Agency. (1983) Final report on the national urban runoff program, Water Planning Division, U.S. EPA, Prepared by Woodward-Clyde Consultants.
5. ASCE. (1999) *National Stormwater Best Management Practices Database, Version 1,* American Society of Civil Engineers, Urban Water Resource Research Council, June.
6. Biemer, Paul, *et al.* (eds.) (1991) *Measurement Errors in Surveys,* New York: John Wiley and Sons.
7. Dillman, D. A. (1978) *Mail and Telephone Surveys,* John Wiley and Sons, New York.
8. Groves, R. M. (1989) *Survey Errors and Survey Costs,* John Wiley and Sons, New York.
9. Lockhart, D. C. (ed.) (1984) *Making Effective Use of Mailed Questionnaires,* Jossey-Bass, San Francisco.
10. Rosenberg, M. (1968) *The Logic of Survey Analysis,* New York Basic Books.
11. Rossi, P., Wright, J., and Anderson, A. (eds.) (1983) *Handbook of Survey Research,* Academic Press, New York.
12. Turner, C.F. and Martin, E. (eds.) (1984) *Surveying Subjective Phenomena,* Vols. 1 and 2, Russell Sage Foundation, New York.
13. Warwick, D. P., and Lininger, A.C. (1975) *The Sample Survey: Theory and Practice,* McGraw-Hill, New York.
14. Sudman S. (1976) *Applied Sampling,* Academic Press, New York.
15. Urbonas, B.R. (1994) Parameters to report with BMP monitoring data, *Proceedings of the Engineering Foundation Conference on Storm Water Monitoring Related Monitoring Needs,* August 7-12, Crested Butte, Colorado, ASCE.

16. Strecker, E. (1994) Constituents and methods for assessing BMPs, *Proceedings of the Engineering Foundation Conference on Stormwater Related Monitoring Needs,* Aug. 7-12, Crested Butte, Colorado, ASCE.

VARIABILITY OF RAINWATER INFILTRATION THROUGH DIFFERENT URBAN SURFACES: *From waterproof top layers to reservoir structures*

G. RAIMBAULT
Laboratoire Central des Ponts et Chaussées (L.C.P.C.),
B.P. 4129, 44341 Bouguenais FRANCE

1. Introduction

Control techniques applied to urban runoff sources often rely upon infiltration. The benefit of such approaches is apparent when recognising the resulting reduction or even elimination of runoff and the concomitant retention of certain pollutants.

Nonetheless, the infiltration of rainwater in urban areas is not necessarily limited to special structures. Various types of urban roadways and surfaces allow, in some instances unintentionally, a portion of runoff to infiltrate. Conventional roadway layers used to pave streets and parking lots are not always as impermeable as road builders would like. The impact of urbanisation creates heterogeneity of an urban area's soil composition, which may lead to flow paths different from those encountered in rural areas. The presence of preferred flow paths can modify the infiltration conditions for reservoir structures (designed to limit urban runoff), thereby intentionally allowing rainwater to infiltrate.

This paper aims to draw a comparison between the hydrological operations of conventional catchments, including standard roadways, and catchments drained by means of special water retention, or reservoir structures. The purpose herein is to provide city managers and public works officials new insights into the relative differences between these two modes of stormwater drainage.

2. Uncontrolled Infiltration in Urbanized Areas

2.1. INFILTRATION AND RUNOFF IN AN URBAN AREA

An initial approach to examining infiltration in an urban setting has been provided through analysing the spatial and temporal variabilities in water volumes discharged during a storm period. In the following, a flow coefficient will be used to represent the ratio of runoff volume, during a given rainfall event, to the total volume of rainwater flowing into the catchment basin.

119

J. Marsalek et al. (eds.), Advances in Urban Stormwater and Agricultural Runoff Source Controls, 119–130.

120

2.1.1. *Temporal Variability of the Flow Coefficient*

Many studies have shown that this flow coefficient varies to a great extent within the same catchment basin and from one rainfall event to the next. An illustration of this finding is provided in Figure 1.

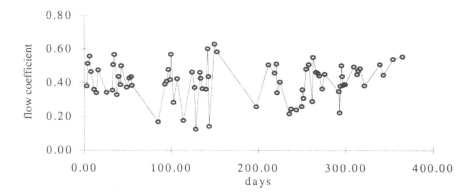

Figure 1. Flow coefficient (under the Batany basin in the Bordeaux area; coefficient of imperviousness: 0.63), [4].

The histograms of flow coefficients observed for two smaller catchments (surface areas 4.7 and 13.4 ha, coefficients of imperviousness 0.37 and 0.39) located at the outskirts of the Nantes metropolitan area show a sizable and similar variability in the flow coefficients, which remain almost always lower than the coefficient of imperviousness and may at times drop to values near 0.

A search for explanatory variables has been performed at the Rezé basin sites [1] using multiple linear regression techniques. Nearly all of the variability explained from such an approach, albeit relatively limited (32% and 38%, respectively), stems from two variables, which in the order of importance are the depth of precipitation during the rainfall event, and the length of the dry period preceding the event. The latter variable serves to characterise catchment wetness.

According to this analysis, the surface areas considered impermeable appear to allow a certain amount of rainwater to infiltrate, which explains, at least in part, the extremely low values of the flow coefficients. Moreover, when the so-called permeable surface areas become saturated with water, they can cause runoff, thereby explaining the occurrence of values higher than the coefficient of imperviousness in some instances.

2.1.2. *Spatial variation of runoff*

The leading cause of spatial variation in runoff is the variability in characteristics of the contributing land parcels. A quick inspection of the various types of roofs has helped draw the distinction between steeply sloped roofs and flat roofs. As for the former, average annual runoff lies in the range of 90% of total rainfall in the more temperate climate zones. For flat roofs, the volume of water retained after drainage can represent in the order of 3% of the volume of gravel placed on such roofs, with this water disappearing primarily through evaporation. The volume of evaporated water therefore is much greater than in the case of steeper roofs. The contribution of steeper-sloped roofs to total catchment runoff depends largely upon the presence of rain gutters; if such is not the case, the majority of precipitation during a rainfall event infiltrates into the ground along the periphery of buildings.

The impermeability of paved roadway surfaces is not always assured. Permeability depends upon the state of the surfacing material and can be roughly estimated using the data in Table 1, [11]:

TABLE 1. Permeability of surfacing materials

State of the surfacing material	Good	Slightly deteriorated or lacking adequate density	Highly deteriorated
Permeability (m/s)	10^{-8}	10^{-7}	10^{-5}
(mm/h)	0.036	0.36	36

Locally specific infiltration can occur along the vertical walls of trench cuts, since the joints between the surface of these cuts and the old surface are rarely watertight. The same often applies to the road pavement and its curbs. Runoff from unpaved urban areas depends primarily upon the degree of compaction of the surface layer and the presence or absence of a "crust". Those zones in which the ground may be saturated potentially contribute, on a temporary basis, runoff.

Role of the geometrical configuration and topography of urban areas. Urban spaces comprise a number of classifications: streets, squares, pedestrian areas, parking lots, building courtyards, school courtyards, parking garage access ramps, etc. Some of these components lie within the public domain, while others are managed privately. Some are laid out in a compact pattern, while others extend in a linear fashion; some are closed, others more open. The conditions for draining off rainwater from these spaces vary from one situation to the next, as do the actual volumes of runoff.

The presence of depressions may lead to a temporary accumulation of water, which leads to partial infiltration. The same type of situation can be encountered in enclosed courtyards. Buildings may serve as obstacles to prevent the drainage of runoff appearing on surfaces located opposite the urban roadway or space. Streets parallel to relief contours impede flow and enhance infiltration.

Streets perpendicular to the contours (with highest slopes) accelerate drainage, thus reducing the potential for infiltration. Both the layout of the urban space and its topography therefore exert an impact on the amount of runoff.

Connection to urban stormwater collection systems. Another factor influencing the drainage rates of urban runoff is the extent of the connection of impervious surfaces to the stormwater collection network.

With respect to residential land uses for example, impervious terraces may discharge runoff onto adjacent lawns. Generally speaking, the water reaching these pervious surfaces infiltrates; however, once saturation has been attained, in the cases with a yard gully or water trap, a portion of the water drained by the network is in fact due to runoff from pervious surfaces.

To conclude this section, the causes of spatial variability in runoff within urban areas are indeed numerous: variability in the component materials of roofs or street / urban space surfaces, effects of topography and the layout of urban spaces, and the extent of direct connection to the sewer collection network.

2.2. PREFERRED FLOW PATHS IN URBAN SOIL SYSTEMS

The previous section has demonstrated that all of the rainwater falling onto a paved area does not necessarily contribute to runoff. This section will discuss what happens to water infiltrating into urban roadways, trench cuts for laying utility lines, and the vicinity of structural foundations. It will serve to specify the role of soil heterogeneity.

2.2.1. *Infiltration into roadways*
Studies focused on better understanding of the wetness state of materials composing roadways and their supporting soil foundation have shown variations in water contents, which are directly related to rainfall conditions and the appearance of "small perched aquifers" at the interfaces between material layers with different levels of permeability [8],[9]. From measurements of water content over a 3-year period, it appears that a fast wetting period distinctly appears during rainfall events followed by relatively slow periods of drainage. The rolling course, which is in good condition, proves representative of many conventional roads throughout France. Statistical analyses have indicated that the most pertinent explanatory variables for the water quantity present in roadways are: the duration of total rainfall during the month preceding the measurements, and the level of evapotranspiration over the same period. The volume of infiltrated water is in essence limited by both the permeability of the surfacing material and the length of rainfall events. An examination of these results shows that annual infiltration through the roadway surface amounts to between 25% and 30% of total rainfall. The infiltration rate into the overlay is in the order of 10^{-7}m/s. The temporary existence of a saturated water level at the interface between the subgrade layer and embankment has been revealed by wet weather flow data from drains installed at this level. These results taken as a whole provide a perspective for the notion of impermeability within the urban environment.

Another impact exerted by the so-called "impermeable" layers is worthy of mention: they suppress or even eliminate the evaporation of water from underlying soil layers. Waterproofing tests were carried out on material storage areas using a bituminous surface dressing spread over unsurfaced ground. The ensuing water concentration measurements showed that the surface dressing was not at all impermeable and moreover that during the rainy season, the wetness of the underlayer increased. During dry periods, the wetness of the ground only dropped very slightly and under all circumstances considerably less than with an unsurfaced ground layer. Urban surfacing materials therefore allow a portion of rainwater to penetrate, yet prevent just about all evaporation.

A method was proposed by Van Ganse [13] to estimate the uniform or localised infiltration within the body of a roadway. His application to pluviometric data for the City of Brussels during the month of November yielded the following results for the same states of surfacing indicated in Section 2.1.2.

TABLE 2. Infiltration within roadways

State of the surfacing material	Good	Slightly deteriorated or lacking adequate density	Highly deteriorated
Percentage of rainwater infiltrating (Brussels in November)	5%	48%	100%

This table demonstrates that as soon as a surfacing material becomes slightly deteriorated, nearly half of the rainwater falling on the surface is able to infiltrate.

This method assumes that the bottom layers of a roadway structure can absorb the infiltrated water volumes. In many instances, permeability contrasts between layers actually cause a localised saturation of the top layer. When this situation arises, the roadway sags may concentrate water circulating longitudinally within the body of the roadway structure, which in the long run can lead to the structural deterioration.

2.2.2. Role of trench cuts with respect to preferred flow patterns within urban soil systems

The role of surface trench cuts for laying utility lines has been highlighted in studies of parasitic waters infiltrating into these networks [2],[6],[5]; this phenomenon is particularly significant for sewer pipes. The additional inflow of water overloads the intake at wastewater treatment plants and is either discharged as untreated effluent or channeled inside the plant for treatment. In the latter case, this additional inflow cannot always be accepted without upsetting the plant operation. The variation in flow rates recorded during the rainy season may be quite large.

An in-depth study of this parasitic water inflow has enabled separating two components (see Figure 2): groundwater drainage (baseflows exhibiting slow variations), and the faster surface drainage flows (occurring over the several days following a rainfall event). The latter component often suggests illicit connections of the stormwater drainage system to the sewer network; in reality, rapid infiltration is apparent when temporary subsurface aquifers are located near the ground surface.

In general, such networks, when not impermeable, function like agricultural drains. Since trenches are often cut below the level of the roadway surface, they may collect water circulating within the body of the road structure, as indicated above, thereby serving as a preferred flow path leading to the eventual outfall (waterproofing fault in the collector system or elsewhere). Different types of trench cuts may intersect, thus creating a situation which complicates any prediction of actual flow patterns. The flooding of cellars in older buildings could be tied to nearby trench excavation for laying new utility lines.

124

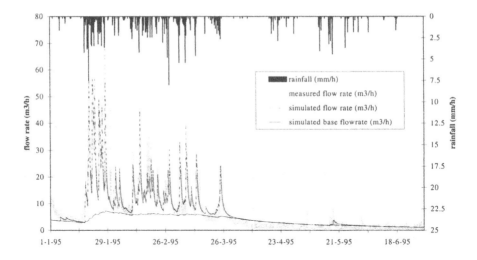

Figure 2. Flow rates of parasitic clean water within a sewer network: Comparison using rainfall data and simulation results

It seems that the phenomenon associated with trench excavations for utility lines can be broadened and applies to all building foundation excavations as well, provided that an outlet has been designed into such work. This likelihood is even stronger for longer and more continuous pipe segments. The author is not aware of any in-depth studies conducted for the case of shallow building foundations. However, such studies for larger and linearly laid out civil engineering projects (e.g. trenches for railroad lines, deeply-embedded trunk sewer lines) have noted a sizable lowering of the groundwater table [12].

These results, along with those obtained within the body of roadway structures, demonstrate that the movement of water in urban undersoil layers is not limited to vertical flows occurring in unsaturated parcels. During the rainy season on many relatively impermeable elements, the trench bottom provides a preferred horizontal flow path within saturated materials, as does the bottom of certain roadway layers.

2.3. MODELLING THE INFILTRATION OF RAINWATER AT THE GROUND-ATMOSPHERE INTERFACE IN SUBURBAN AREAS

The weakest feature of urban hydrological models pertains to the generation of runoff. In reality, such models have often been designed for sizing a stormwater drainage network and thus implicitly consider that the contribution of a catchment basin depends primarily on the level of imperviousness. The response of a given catchment is related to the state of saturation of both the surface cover and the undersoil, as well as to the nature of the undersoil. A third dimension (vertical mechanism) is in fact required in this type of models and that is one of the objectives of the work described below.

This section describes the simulation of the infiltration occurring in a vertical plane perpendicular to the road network representative of a suburban residential environment [1]. The ground has been modelled using a finite element computation code developed for either saturated or unsaturated media. At the interface between the atmosphere and the various elements (building, unsurfaced ground, roadway), reservoirs representing the interception of runoff have been placed. This modelling technique assumes a certain knowledge of the hydrodynamic properties of both the ground and the roadway. The roof of the building is assumed to be impermeable, yet it allows for a small amount of runoff interception. The permeability used for the surface of the roadway corresponds, for the given scenario, to a surfacing material that is slightly deteriorated or of low density.

Analysis of the findings from 7 consecutive years of observation in the west of France for the various types of urban surfaces has led to the results presented in Figure 3, which indicate the percentages of evaporation, runoff and infiltration with respect to the quantity of rainfall. Over a surface area situated between the building and the roads/urban spaces, the percentage of evapotranspiration above 100% is due to water transfer taking place beneath the roadway and the neighbouring building. A comparison of the results obtained displays a variance in hydrological behaviour. Special attention should be paid to the high level of infiltration in the roadway, which provides support for the estimations proposed by Van Ganse (see Section 2.1.1).

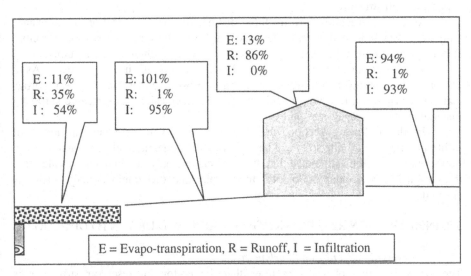

Figure 3. Estimation of the evapotranspiration, runoff and infiltration for four types of suburban road surfaces (percentages with respect to the quantity of rainfall).

An analysis of the infiltration on an unpaved (or "natural") parcel of land located between the building and roadway, for a winter rain event (during the wet season), has led to the following results. At the beginning of the event, infiltration is equal to the amount of precipitation, but afterwards infiltration becomes very small. This evolution is tied to developments of the saturated ground surfaces. Moreover, it can be observed that the building and roadway both exert an influence on this development. Underneath these structural elements, which are impermeable or at least less permeable than the surface of the ground, the soil remains unsaturated and can still store additional water.

3.   Reservoir Structure and Controlled Infiltration

3.1. INTRODUCTION

The higher quantity of runoff in urban areas is largely due to continual increases in the surface area occupied by urban roadways. From the premise of using these areas to mitigate runoff and its consequences in terms of pollutants discharged into the receiving waters, the concept of the reservoir structure has been derived. The storage of a relatively low water depth over large surface areas provides for a sizable water retention volume and can facilitate infiltration into the soil layers. The term "reservoir structure" reflects the two functions performed by the component porous materials, i.e.: **a mechanical function** (associated with the word "structure"), which pertains to bearing variable loads, from pedestrian traffic in pedestrian areas to automobile traffic on roadways; and a **hydraulic function** (associated with the word "reservoir"), which is performed through the porosity of the component materials and enables temporary retention of water prior to its disposal, if possible by means of infiltration into the supporting soil structure.

Reservoir structures perform three distinct functions: collection of rainwater, temporary retention of this rainwater, and disposal. The type of reservoir structure selected depends upon the mode of execution chosen for each of these functions.

Rainwater infiltration into the soil can only occur in two cases. As a precondition, the infiltrated water must have crossed either the coarsest porous material serving as a reservoir, or this material is preceded by a permeable surfacing material containing small-size pores.

In the following section, the effects of reservoir structures in terms of infiltration into underlying soils will be discussed both quantitatively and qualitatively. Furthermore, it will be stipulated that these effects can only be obtained provided that the design has been adapted to local topography and that minimal maintenance is ensured.

3.2. INFILTRATION RESERVOIR STRUCTURES AND URBAN HYDROLOGY

3.2.1.   *Reservoir structures and urban layout*
Temporary retention of water requires that the bottom of reservoir structures is relatively flat. If such were not the case, water would accumulate in the lower part and the subsequent rise in water height could cause overflows. Whenever the sequence of planned structural elements extends linearly (e.g., along a street), it is preferable to lay out such elements parallel to contour lines. In this way, the city can be decomposed into runoff zones separated at regular intervals by infiltration structures. When such an arrangement is impossible, a decomposition into a series of successive retention basins, using waterproof separating walls, is required.

This perspective helps convey the idea that in order to facilitate infiltration in urban areas through reservoir structures or other techniques, such an objective must be firmly incorporated into the hydrological design at the time of adoption of the initial layout. The various partners in the urban facilities project must therefore be kept informed about and involved in the intended design approach.

Moreover, a succession of runoff and infiltration zones can be arranged in the same way as encountered in natural settings (e.g. in Niger's bush, a natural ecosystem is organized into strips along contour lines, alternating surface runoff areas and highly-infiltrating dense forestland). In examining the downstream contributions of a catchment treated in this manner, the observation can be drawn that the "contributing zones" remain limited and quite distinct from the "active zones" over which runoff is generated.

3.2.2. *Hydrological operations of infiltration reservoir structures*

The monitoring of a roadway equipped with a reservoir structure built on relatively impermeable soils has shown that even in this case it is possible for a very high percentage of rainwater to infiltrate (see Table 3). This experiment was in fact unfavourable to cause infiltration, since no mechanism had been set up to limit flows at the outfall, located at the base of the structure on ground level. Such a mechanism could have served to retain the water from major rainfall events for longer periods and to increase infiltration.

TABLE 3. Values, expressed in terms of per-rainfall event infiltration

Events	Infiltration/rainfall duration (mm/hr)		Infiltration/water retention period (10^{-7}m/s)	% infiltrated
- with lateral discharge	Minimum	0.5	0.5	87
(at the base of the structure)	Maximum	3.6	16.9	> 87
- without lateral discharge	Maximum	5.8	19.8	100

(In this table, infiltration is assumed to equal the difference between the quantity of rainfall and the amount of flow discharged at the base of the roadway. The effect of evaporation is presumed to be weak).

After an initial assessment, it would appear that the infiltration flow rate varies considerably from one rainfall event to the next as a result of variations in soil wetness conditions. Simulating this infiltration is possible through the use of flow-based computation codes for unsaturated media. Nonetheless, the coupling taking place between the reservoir structure and the soil tends to be complicated by both the potential discharge occurring at the base of the roadway and any heterogeneities existing in the soil (e.g. trench cuts for laying utility lines).

3.2.3. *Improvement in the quality of infiltration water through reservoir structures*

The evolution in water quality can be analyzed from two perspectives: inside the reservoir structure itself, and in the underlying soil. Table 4 presents the level of pollution reduction taking place within the structure at various sites in France.

Table 4. Pollution reduction by infiltration into a roadway equipped with a reservoir structure

Site	Characteristics	Pollution reduction (% concentration)				
		TSS	COD	Pb	Zn	BOD
Rezé (Nantes)	drainage overlay + porous untreated gravel	61		81	67	
Parc Caillou (Bordeaux)	drainage overlay + porous concrete	36	79	86		
ZONE I Verneuil –(Paris)	drainage overlay + porous untreated gravel	81	63	76	35	45
ZONE II Verneuil –(Paris)	conventional overlay + porous untreated gravel	68	48	77	45	39
ZONE III Verneuil –(Paris)	group of various reservoir structures	1	14	50	16	7

With the exception of the results from Verneuil in Zone III, which seem to have been influenced by a public works project carried out on the site, *this table indicates that drainage of runoff into reservoir structures serves to improve water quality, regardless of the permeability of the surfacing material. However, the presence of a drainage-enhancing cover strengthens this effect.*

Pollutants are primarily trapped in the upper part of the surface layer in the case of a permeable cover material. The way in which these pollutants are then handled depends upon the maintenance effort aimed at avoiding clogging (see below). The evolution in both water and soil quality underneath the reservoir structure could be assessed from a number of soil samples and numerical simulations. Soil analysis conducted at the Rezé site, 4 and 8 years after construction, has not revealed any significant evolution [3]. The comparison of concentrations of heavy metals contained in the soil under conventional roadways (without reservoir structures) in the Bordeaux region has not highlighted any major discrepancies either.

In order to estimate the concentration of micro-pollutants in the soil and the long-term infiltration water, a simulation was run with the "LEACHM" model for a 50-year period, assuming that the water at the base of the reservoir structure could be more heavily polluted than the values observed (initial values: Pb: 35 μg/l, Cu: 22 μg/l, Cd: 3.5 μg/l, and Zn: 340 μg/l). When inputting a particularly unfavourable data set, the simulations indicated that the concentrations of lead diminish rather quickly as a function of depth, whereas those of cadmium decrease less quickly due to the soil's lower retention rate for this metal. The concentrations become low as of a depth of 30 cm. The risk of groundwater contamination of deeper aquifers is therefore rather remote [7].

3.2.4. *Clogging and maintenance*

The operations of reservoir structures with a permeable surfacing material depend upon the ability to preserve this permeability over time. Recent studies [10] have shown that clogging materials can be divided into two components: a sandy primary sealant and a small quantity of a finer secondary sealant (clay content accounts for < 2% of total mass). The evolution of this clogging seems tied to the succession of wetting-drying cycles corresponding to the alternation of rainy and dry weather periods. A study conducted with a model simulating the sandy fraction of the clogging material has made it possible to analyze the bonds formed by clay in between the larger grains. This study revealed the presence of "veils" capable of obstructing pores with a small quantity of clay. The most prudent method for preventing the surfacing material from clogging is to regularly draw off the primary sealant before the clay adds too much cohesion to the surfacing material.

Sealants play an important role in trapping micro-pollutants. The concentration of heavy metals turns out to be high only within the first few centimetres of drainage-enhancing road covers. Maintenance techniques using suction or high-pressure scrubbing + suction (should preventive cleaning be insufficient) do generate polluted products. In order to reduce the quantity of such waste products, studies are being conducted on methods for separating the coarser, relatively unpolluted material from the finer, more heavily-polluted material.

4. Conclusion

While the infiltration of rainwater in an urban setting is less significant than that in rural areas, it proves to be highly variable both spatially and temporally. Some urban roads and surfaces may allow up to 50% of the total annual rainfall received to infiltrate. The subsequent flow course of this infiltrated water is not well known; the flow pattern is complex, particularly due to the variability in types of soil surfaces, urban layout and heterogeneity of the undersoil as a result of the manipulations taking place at this layer. It is commonly necessary to account for the horizontal subsurface flows in urban soils. Infiltration therefore is not simply a vertical flow in an unsaturated soil. Unintended drainage systems are often present as well. At this stage, further understanding of the ultimate course of infiltration water in urban areas would be most helpful. Infiltrated water influences not only the mechanical behaviour of foundation soils of buildings and other structures, but also the potential replenishment of groundwater sources. It can upset successful operations of such public works as sewer collection and treatment systems.

A system for draining stormwater at the source, which enhances infiltration by such infrastructures as reservoir structure-equipped roadways, enables to control infiltration. Such a system incorporates infiltration into the design of urban public works projects and identifies the means for achieving optimal infiltration. It causes a sizable reduction in, if not the entire elimination of, urban runoff, while avoiding the discharge of pollutants at the outfall during storms and stimulates the replenishment of groundwater sources. This approach to urban stormwater drainage does necessitate however a preliminary in-depth assessment of rainwater infiltration and its control.

5. References

1. COUREUL C. (1998) Evaluation d'une modélisation à base physique de la transformation pluie-débit en milieu urbain, Joint "DEA" dissertation INA P-G, Paris VI and ENS, 45 p.
2. BERTHIER E. (1999) *Contribution à une modélisation hydrologique à base physique en milieu urbain*, Doctoral thesis, Institut National Polytechnique de Grenoble, 196 p.

3. SETRA, LCPC (1992) *Guide technique : Ecrans drainants en rives de chaussées,* SETRA, Bagneux, 71 p.
4. RAIMBAULT G. (1986) Cycles annuels d'humidité dans une chaussée souple et son support, *Bull. Liaison Labo. P. et Ch.* 145 (Sept.-Oct.), 79-84.
5. RAIMBAULT G., SILVESTRE P. (1990) Analyse des variations d'état hydrique dans les chaussées, *Bull. Liaison Labo. P. et Ch.* 167 (May-June), 77-84.
6. VAN GANSE (1978) *Les infiltrations dans les chaussées : Evaluations prévisionnelles,* in Symposium Report on Road Drainage, OECD, Bern, 1978, pp. 176-192.
7. BREIL P., JOANNIS C., RAIMBAULT G., BRISSAUD F., DESBORDES M. (1993) Drainage des eaux claires parasites par les réseaux souterrains : De l'observation à l'élaboration d'un modèle prototype, *La Houille Blanche* 1, 45-57.
8. JOANNIS C., BELHADJ N., RAIMBAULT G. (1994) *Le drainage d'eaux claires parasites par les réseaux d'assainissement en période pluvieuse,* in Hydrotop, Marseille, April 12-15. International scientific and technical symposium "Mieux gérer l'eau".
9. DUPASQUIER B. (1999) *Modélisation hydrologique et hydraulique des infiltrations d'eaux parasites dans les réseaux séparatifs d'eaux usées,* Doctoral thesis, Ecole Nationale du Génie Rural, des Eaux et Forêts, Paris, 209 p.
10. TOMACHOT M. (1986) *Impact d'un aménagement urbain sur une nappe aquifère: Cas de la ville nouvelle d'Evry (Essonne),* S.H.F. XIXth Days of Hydraulics, Paris, Sept., P. H-3-1-II-3-8.
11. COLANDINI V. (1997) *Effets des structures réservoirs à revêtement poreux sur les eaux pluviales : Qualité des eaux et devenir des métaux lourds,* University of Pau thesis, January 1997, 171 p. + appendices.
12. LEGRET M., NICOLLET M., MILODA P., COLANDINI V., RAIMBAULT G. (1999) Simulation of heavy metal pollution from stormwater infiltration through a porous pavement with reservoir structure, *Wat. Sci. Tech.* 39, 119-125.
13. RAIMBAULT G., NADJI D., GAUTHIER C. (1999) *Stormwater infiltration and porous material clogging,* in I.B. Joliffe and J.E. Ball (eds.), Proc. 8 ICUSD, Sydney, August 30 - Sept. 3, 1999, pp. 1016-1024.

FIELD-INVESTIGATIONS OF POLLUTANTS IN STORMWATER RUNOFF, SEEPAGE WATER AND TOPSOIL OF STORMWATER INFILTRATION SITES

F. REMMLER and U. HÜTTER
Institut für Wasserforschung GmbH Dortmund
Zum Kellerbach 46
D-58239 Schwerte

1. Introduction

Nowadays, the natural water cycle is fundamentally influenced by anthropogenic factors. In recent years, increasing urban settlement in drainage areas has led to intensified surface paving and consequently to an enormous increase in the water to be collected and transported from urban areas. Therefore, current approaches to wastewater collection have caused different economic and environmental problems in urban hydrology. The main problems are:
-   overloading of sewage treatment plants,
-   increasing number of flood events,
-   reduced groundwater recharge,
-   adverse impacts on the quality of receiving water by, e.g., combined sewer overflows, and
-   increasing investment and operating costs for sewage treatment facilities.

For these reasons, water engineers have been searching for an alternative to the current approach to urban drainage systems via separate and combined sewers. A possible solution is the management of stormwater runoff as an alternative to conventional drainage systems. This management procedure uses the main principles of drainage, retention and infiltration, both separate and in various combinations, depending on the porosity of the soil and the availability of open land. In many cases the infiltration of stormwater runoff is already practised. This kind of infiltration management increases the input of water per infiltration area unit (m^2) as well as the volume and the spectrum of substances. With the aim of achieving soil and water protection, it is therefore necessary to question whether, and in which cases, stormwater infiltration is sustainable and environmentally sound.

The Institute for Water Research has done extensive investigations into the impact of stormwater infiltration on the quality of soil and seepage water. In several research projects at different locations, the effect on environmental components was monitored over a period of 2 to 5 years. This paper presents selected results of investigations at different infiltration troughs.

J. Marsalek et al. (eds.), Advances in Urban Stormwater and Agricultural Runoff Source Controls, 131–140.

2. Potential Pollution of Stormwater Runoff

In passing through the atmosphere, precipitation accumulates many chemical substances. The spectrum of polluting and dangerous substances normally increases during the flowpath from the catchment area to the drainage system. The pollution of precipitation and stormwater runoff is a result of the ubiquitous distribution of pollutants in the environment and the deposition of pollutants on all elements of the drainage area.

For more than 30 years many research projects have investigated the quality of precipitation and the different types of surface runoff (e.g. roof runoff, street runoff). The extensive results give evidence for varying pollution depending both on the local conditions and on the analytical conditions of the specific investigation. Therefore, these isolated cases are not universally applicable and cannot be transferred to other types of catchment areas.

At the moment, the different types of surface runoff components can be classified only with simplistic approaches. The draft of the new German A 138 Standard of the Abwassertechnische Vereinigung (ATV) [1] arranges in a decision-matrix some typical surfaces in quality categories according to their expected potential degree of pollution and the possible infiltration systems (Table 1).

The granting of permission to infiltrate depends on the quality of the stormwater runoff and the kind of treatment facilities. Unpolluted and less polluted water can be reintroduced immediately into the natural water cycle, if some requirements related to the location, the structure and the operation of the infiltration system are fulfilled [1, 2, 3, 4].

For example, as a rule the runoff from surface types 1 and 2 in Table 1 can be infiltrated with all systems. Infiltration systems which do not use passage through the topsoil are permissible only in special cases (e.g. lack of open ground) for runoff from surface types 1 to 7 if the connected surfaces are outside of areas with high atmospheric emissions. For runoff from surface types 3 to 11, only decentral and central infiltration systems with the passage through the topsoil are allowed. Depending on the pollution level for some types of runoff, treatment steps before infiltration are required. In general - with only a few exceptions - surface runoff from surface types 12 and 13 is not allowed to be infiltrated.

The investigations which have been carried out to date have identified many chemical substances in stormwater runoff. The following micropollutants deserve particular attention, due to their toxicity, persistence and accumulation in biological systems:

- organic micropollutants (e.g. polycyclic aromatic hydrocarbons, PAHs), and
- inorganic micropollutants (e.g. heavy metals).

PAHs and some heavy metals are transported predominantly through their association with suspended solids. An assessment of the quality of stormwater runoff is difficult because there are no special quality standards.

During infiltration, different processes cause a significant reduction in some pollutants. However, physical or chemical adsorption and biological processes lead to increased concentrations of pollutants in the soil of such infiltration systems.

Biological degradation of some kinds of pollutants (e.g. heavy metals) in natural systems is negligible or does not occur at all. Therefore, an accumulation of chemical substances in the topsoil can exhaust the buffer and filtration ability of the soil. The next section presents some results of investigations concerning this aspect.

TABLE 1. Stormwater infiltration in consideration of the characteristics of the runoff producing surfaces (A 138 [1], revision: state of discussion August/2000).

Type of surface	Increase in pollutant load	Assessment of quality	surface infiltration systems				subsurface infiltration systems	
			infiltration area	infiltration trough, trough-trench-element	central infiltration basins with Ared : As < 15:1	central infiltration basins with Ared : As > 15:1	trench or pipe elements	shaft infiltration
1	2	3	4	5	6	7	8	9
1 green roofs		harmless	+	+	+	+	+	+
2 terraces and non metallic roofs in residential and comparable business areas			+	+	+	+	+	(+)
3 bikeways and sidewalks in residential areas, areas with reduced traffic			+	+	+	(+)	(-)	(-)
4 yards in residential and comparable business areas			+	+	+	(+)	(-)	-
5 parking areas with less traffic density			+	+	+	(+)	(-)	-
6 streets with less than 2000 cars/day,			+	+	+	(+)	(-)	-
take-off runways, landing strips and runways of low-traffic airports, runways of heavy-traffic airports			+	+	+	(+)	-	-
7 non metallic roofs in other business and industrial areas		tolerable	+	+	(+)	(+)	(-)	-
8 streets with more than 2000 and less than 15000 cars/day, take-off runways and landing strips of airports			+	+	(+)	(+)	-	-
9 parking areas with high traffic density			+	(+)	(+)	(+)	-	-
10 metallic roofs, agricultural yardgrounds			+	(+)	(+)	(+)	-	-
11 streets with more than 15000 cars/day [1]			+	(+)	(+)	(+)	-	-
12 yards and streets in other business and industrial areas		not tolerable	(-)	(-)	(-)	(-)	-	-
13 special areas, e.g. truck parking areas and lots, airplane position areas of airports			-	-	-	-	-	-

+ as a rule allowed   (-) allowed in exceptional case

(+) as a rule allowed as long as cleaning is possible   - not allowed   [1] thickness of zone of percolation at least 2 m

3. Investigations at Different Infiltration Sites

For monitoring the development of the water quality of the precipitation and the runoff percolating through the topsoil of the trough and the trench, measurement networks for

134

water samples were installed at two study sites, in Dortmund and Gelsenkirchen. Water samples were collected there over a period of 2 to 5 years.

For estimating potential accumulation effects of chemical substances in alternative drainage systems, it was necessary to observe the topsoil in several infiltration troughs. Another significant research aspect was whether long-term operation of the trough-trench system could exhaust the buffer and filtration capacity of the soil.

For these reasons, different drainage systems in Germany have been investigated with respect to the quality of soil. Soil samples were taken from the infiltration troughs at several depths after different lengths of time. At new infiltration sites, these samples were compared with a control sample taken during the construction work. At older infiltration sites, soil samples next to the infiltration areas were taken to estimate the accumulation effects.

3.1. STUDY SITES AND BACKGROUND INFORMATION

The first investigation site is located in the grounds of a daycare centre in the urban area of Dortmund in North-Rhine-Westphalia. In 1993 the owner of the building, the Dortmunder Stadtwerke AG, chose a trough-trench system as the stormwater drainage method for the building. An approach such as this allows for the decentralised retention of stormwater and the infiltration of stormwater through a system of infiltration troughs with a trench underneath. The outdoor grounds, including the trough-trench system, were finished in autumn 1994. The drainage design consists of three grass-surfaced troughs and a trench below filled with lava rock. The groundwater level can be found at about 15 m below the trench. The trough-trench system receives runoff mainly from the roofs, the terraces and a small car park. In total, nearly 1,030 m^2 of drainage surface are connected to the drainage system [5]. For the investigated infiltration trough, the ratio between the connected impervious surfaces (A_{red}) and the infiltration area (A_S) is 5.5 to 1 (A_{red}/A_S).

The second investigation site is located in the urban settlement of Gelsenkirchen-Schüngelberg in North-Rhine-Westphalia. There are also several trough-trench systems used as the stormwater drainage method for the runoff of roofs and terraces. The systems were built between 1992 and 1996. For the investigated infiltration troughs, the ratio between the connected impervious surfaces (A_{red}) and the infiltration area (A_S) is 6 to 1 (A_{red}/A_S).

Another investigation site is located in the suburb of Frohnau in the city of Berlin. There is an old infiltration trough for stormwater runoff with an estimated operation time of 50 years at Fürstendamm Street. Because the time of the last sludge clearance is uncertain, a maximum of 50 years time in operation is assumed. This central trough receives the runoff mainly from road areas with a low traffic frequency. The investigated infiltration trough has a ratio between the connected impervious surfaces and the infiltration area of 6.7 to 1 (A_{red}/A_S).

3.2. INVESTIGATION METHODS

At the investigation sites in Dortmund and Gelsenkirchen, water samples were taken after precipitation and runoff events at special monitoring networks. The monitoring stations were:

(a) precipitation

(b) roof runoff (= input into the infiltration systems)

(c) seepage water after 30 cm topsoil passage

(d) discharge of the trench below the trough

The soil samples were collected near to the influx area of the troughs at the following depths: 0 - 5 cm, 5 - 10 cm, 10 - 20 cm and 20 - 30 cm. The soil samples at the troughs were collected and investigated after different times of operation. To compare the results, control samples were collected during the construction of the systems or at the older systems in Berlin outside of the infiltration area.

In order to determine the total amount of substances (e.g. heavy metals) and their leachability by changing pH-conditions, different examination methods were used:

(a) determination of the acid soluble portion of metals (DIN 38414/7 S7)[6],

(b) determination of leachability by water (DIN 38414/4 S4)[7], and

(c) determination of leachability by water with pH 4 (pH_{stat} leaching test) [8].

Whereas the S4 procedure (b) only represents the starting conditions of the leaching processes in nature, the pH_{stat} - procedure (c) takes greater consideration of the long-term development of the pH-value in nature. The pH_{stat} - procedure also gives information about the neutralisation buffer capacity of the soil for acids and bases. These characteristic magnitudes that describe the long-term leaching behaviour, in connection with data about the respective soil (coefficient of hydraulic conductivity, stratum thickness), yield conclusions about the probability of the worst case of leaching occurring.

3.3. SELECTED RESULTS

The mean concentrations (sampling period of 2.5 years) of selected substances in Table 2 illustrate the quality development ranging from the precipitation and the roof runoff to the discharge of the trench beneath the troughs at the infiltration sites in Dortmund and Gelsenkirchen. The atmospheric inputs of heavy metals and PAHs are comparable in Dortmund and Gelsenkirchen. The roof runoff absorbs especially a lot of zinc from the materials of the roof drainage system, so that the concentrations of zinc often reach high values. The zinc concentrations of the roof runoff in Gelsenkirchen are much higher than in Dortmund because at this site the first flush of the roof runoff was sampled and the materials of the roof drainage system are much older (higher corrosion effects). The low zinc concentrations of the seepage water after the topsoil passage show the good treatment effect of the topsoil. There are no significant differences between the concentrations of the seepage water and the trench water. The concentrations of the heavy metals in the seepage water and the trench water were always significantly below the audit values of the German federal soil protection decree [9].

136

TABLE 2. Mean concentrations of selected substances in water samples (precipitation, roof runoff, seepage water after topsoil passage, trench water) at stormwater infiltration sites in Dortmund and Gelsenkirchen.

Substance		Precipitation		Roof runoff		Seepage water after 30 cm topsoil passage		Trench water		Audit value BBodSchV * [9]
		Dort-mund	Gelsen-kirchen	Dort-mund	Gelsen-kirchen	Dort-mund	Gelsen-kirchen	Dort-mund	Gelsen-kirchen	
Cadmium**	[µg/l]	2,2	0,6	0,4	0,5	0,1	n. s.	0,2	0,2	5
Copper**	[µg/l]	10	8	24	10	8	n. s.	19	12	50
Lead**	[µg/l]	9,6	5	1,4	2	1,1	n. s.	2,9	1,2	25
Zinc**	[µg/l]	59	73	571	4266	18	n. s.	18	48	500
FLA	[ng/l]	85	109	85	24	30	n. s.	23	18	n. i.
NAP	[ng/l]	41	74	60	28	19	n. s.	23	18	2000
PHE	[ng/l]	79	90	88	23	14	n. s.	15	9	n. i.
PYR	[ng/l]	61	72	60	18	24	n. s.	20	16	n. i.

* = result track soil - groundwater, ** = filtrated, n. s. = no sample, n. i. = no information
FLA = Fluoranthene, NAP = Naphthalene, PHE = Phenanthrene, PYR = Pyrene

The concentrations of PAHs fluoranthene, naphthalene, phenanthrene and pyrene in the seepage water or trench water were compared with the precipitation and the roof runoff, and showed the same good treatment effect resulting from percolation through the topsoil (Table 2). In Gelsenkirchen, there is already a reduction of the concentrations in the roof runoff because the high amount of organic material from nearby trees and plants in the roof drainage system intensifies sorption processes.

Figures 1 and 2 show, for the sites in Gelsenkirchen and Berlin, the acid soluble amount of zinc and the proportion of zinc which can be mobilised by the pH_{stat} leaching test with pH 4 in the topsoil of the infiltration troughs after 39 months and 50 years of operation. The pH_{stat} leaching test with pH 4 describes the worst case of leaching because this low pH probably facilitates the release of most soluble heavy metals from a soil. In addition, a sample comparison is presented between the topsoil of the trough and the control mple that was taken in Gelsenkirchen during the construction work. For the system in Berlin, zinc concentrations from soil samples taken outside the trough are given as reference samples.

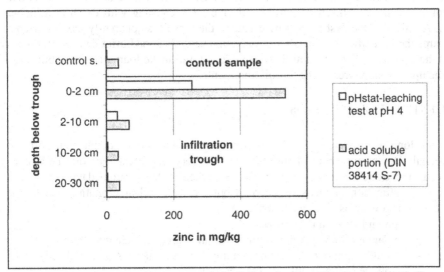

Figure 1. Acid soluble and mobilised amount of zinc in the soil of an infiltration trough with 39 month operation time in Gelsenkirchen.

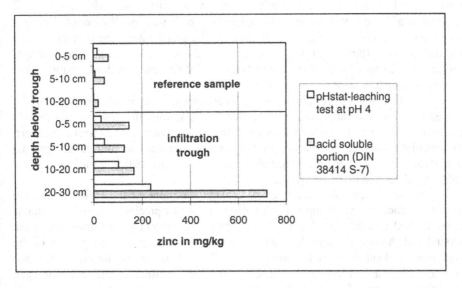

Figure 2. Acid soluble and mobilised amount of zinc in the soil of an infiltration trough with 50 years operation time in Berlin.

The investigations with both methods for Gelsenkirchen showed, in comparison with the control sample, a high accumulation of zinc in the top 2 cm. This can be explained by the fact that zinc is a major part of the materials that were used for the gutter and roof drain pipes. The results also showed a slight decrease of the pH-value and the neutralisation buffer capacity of the soil for acids with time.

Figure 2 shows the results for zinc in the topsoil of the old infiltration trough

138

in Berlin. In contrast to the results for the system in Gelsenkirchen, the acid soluble amount and the mobilised proportions of zinc are increasing with depth at a higher rate. At this site, the first immobilised zinc in the topsoil is apparently moving deeper towards the groundwater. The almost exhausted neutralisation buffer capacity for acids and the very low pH of the soil (between 4.2 and 4.8) make long-term immobilisation in the upper centimetres of the topsoil impossible.

3.4. ESTIMATION OF LOADS

For the long term operation of infiltration systems, knowledge of the transport and especially the whereabouts of the pollutants is very important. For one part of an investigated trough trench system in Gelsenkirchen with measured data over 2 to 5 years, an estimation of loads was carried out. For selected precipitation events, the following data were used for a load balance:
-   precipitation in mm (measured),
-   volume of precipitation of the infiltration area A_s (calculated),
-   trough input over the connected impervious surfaces A_{red} (simulated), and
-   volume of the throttle outlet (measured).

The simulation of the trough input was done with the hydrological storage model MURITEST by the Institut für Wasserwirtschaft, Hydrologie und landwirtschaftlichen Wasserbau of the University of Hannover. For selected events, the measured unfiltered metal concentrations of the precipitation, the roof runoff and the trench water (throttle outlet) were used for the load calculation. These sample results are only spot samples of the substances and their concentrations, and therefore the load calculation must be seen as an estimation.

For the calculation of the volumes, it was assumed that the trough volume input passes the topsoil without loss into the trench and is then divided into the throttle discharge and the part which infiltrates into the underground. The difference between the calculated input of substances by the precipitation and the roof runoff and the output of substances by the infiltration and the throttle outlet is defined as the substance detention in the trough soil.

Figure 3 shows the calculated percentage variations of the system input and output for zinc. The whole input in this balance caused by precipitation and roof runoff is set as 100 percent. The losses of zinc out of the system by infiltration into the ground and through the throttle outlet are very low, because over 95 percent of the zinc input is kept back in the trough soil. The volume of the throttle outlet and consequently the outflow of substances out of the system is different for each precipitation event. The start, the duration and the volume of an overflow event depend on many conditions (e.g. precipitation depth and intensity, water level in the trench).

The results of the corresponding investigations of the zinc content in the topsoil of the trough confirmed the load estimation and the high detention of zinc in the trough. Therefore, at this infiltration site, an average accumulation of about 34 mg zinc/kg topsoil per year is estimated.

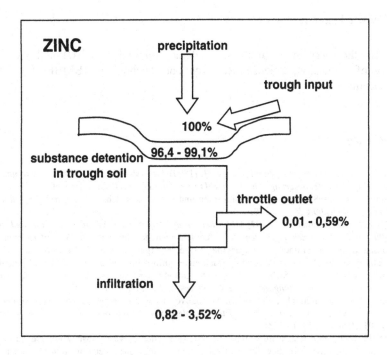

Figure 3. Percentage variations of the in- and output components of zinc in an investigated part of a trough-trench-system in Gelsenkirchen.

4. Conclusions

Experiences with the planning and construction of several stormwater infiltration facilities and the soil investigations at some infiltration sites with shorter and longer times of operation showed that it is absolutely necessary to define specific criteria for stormwater runoff infiltration as a form of stormwater management. For long-term protection of soil and groundwater, it is important to follow specific guidelines for the planning, construction and operation of stormwater infiltration sites.

The risk of groundwater pollution can be minimised by using infiltration systems with low hydraulic surcharge, in which the stormwater runoff percolates through the topsoil. An ecologically tenable stormwater runoff infiltration system can be achieved only by exploitation of the treatment and adsorption potential of existing and developing infiltration technologies and their combinations. This will only function if, in addition, the requirements for the location, structure and operation are fulfilled.

Furthermore, there is a need to develop and use suitable instruments to monitor the long-term effectiveness of the infiltration systems. After long periods in operation, proper cleaning and disposal methods for the substances accumulated in the topsoil of infiltration systems must be used.

140

5. Acknowledgement

We thank the German Environment Federal Foundation (DBU) and the Federal Ministry of Education, Science, Research and Technology (BMBF) for financial support of this work.

6. References

1. ATV - Abwassertechnische Vereinigung e. V. (1999) *Arbeitsblatt A 138 - Planung, Bau und Betrieb von Anlagen zur Versickerung von Niederschlagswasser*, Gelbdruck, GFA, Hennef.
2. Remmler, F. (1998) Regenwasserversickerung und Grundwasserschutz, *Zeitschrift für Kulturtechnik und Landentwicklung* **39**, 272-279.
3. Hütter, U. und Remmler, F. (1997) *Möglichkeiten und Grenzen der Versickerung von Niederschlagsabflüssen in Wasserschutzgebieten*, Veröffentlichungen des Instituts für Wasserforschung GmbH Dortmund und der Dortmunder Energie- und Wasserversorgung GmbH, Nr. 54, Dortmund.
4. Remmler, F. und Schöttler, U. (1998) Qualitative Anforderungen an eine naturnahe Regenwasserbewirtschaftung aus der Sicht des Boden- und Grundwasserschutzes, in F. Sieker (ed.), *Naturnahe Regenwasserbewirtschaftung*, Analytica Verlagsgesellschaft, Berlin, 104 - 125.
5. Stecker, A. and Remmler, F. (1996) Alternative stormwater drainage concept and design - a demonstration object -, *Proceedings of the 7th International Conference on Urban Storm Drainage*, Vol. III, Hannover, 1831 - 1836.
6. DIN 38414/7. German standard methods for the examination of water, waste water and sludge; sludge and sediments (group S); digestion with aqua regia for subsequent determination of the acid soluble portion of metals (S7).
7. DIN 38414/4. German standard methods for the examination of water, waste water and sludge; sludge and sediments (group S); determination of the leachability by water (S4).
8. Obermann, P. und Cremer, S. (1993) *Mobilisierung von Schwermetallen in Porenwässern von belasteten Böden und Deponien: Entwicklung eines aussagekräftigen Elutionsverfahrens*, Landesamt für Wasser und Abfall (Hrsg.), Essen. (=Materialien zur Ermittlung und Sanierung von Altlasten, Band 6).
9. Bundes-Bodenschutz- und Altlastenverordnung (BBodSchV) (12.07.1999), in Bundesgesetzblatt Jahrgang 1999 Teil I Nr. 36, Bonn, 1554 - 1579.

CONSIDERATIONS FOR THE FIRST APPLICATION OF SOURCE CONTROL MEASURES FOR STORMWATER RUNOFF IN THE ATHENS METROPOLITAN AREA

E. AFTIAS
National Technical University of Athens
5, Heroon Polytehniou Str., 157 80 Athens, Greece

1. Athens Hydrographic Network

The Greater Athens area consists of two main catchments, the catchments of the rivers Kifisos and Ilisos, which before the urbanisation, occurring during the last 50 years, possessed dense hydrographic networks. Both rivers drain to the Saronic gulf and their estuaries are only 1.5 km from each other. The main reach of the Kifisos has a total length of 21 km, draining an area of 390 km². The urban part of the catchment is continuously increasing, currently amounting to 60% of the total area. It is also important to point out that during recent years, extensive summer fires have drastically changed the behaviour of the non-urban part of the catchment. Over a length of 8 km upstream from its outlet, the Kifisos river is flowing in a closed section, with a capacity of 1400 m³/s corresponding to a 50-year flow. Upstream from this section, the Kifisos has a variable open cross-section of structural concrete, with dimensions corresponding to various levels of flow, with some levels as low as 5 years.

Along the river, a number of illicit connections continue to discharge industrial and domestic sewage into the river. During dry weather periods, baseflow and sewage are diverted 5 km upstream of the outlet to an adjacent sewage collector, which conveys them to the sewage treatment plant of Psitalia. During storm events, the river transports to the sea an important pollution load from domestic and industrial illicit connections already mentioned, but above all, from the wash-off of streets and the rest of urban surfaces drained. Taking into account the pertinent climatic conditions characterised by a long dry summer period, the first significant autumn rains have a shock effect on the sea environment.

Next to the Kifisos, the Ilisos drains the east and northeast areas with a catchment of 52 km², mostly urban. The length of the main reach is 15 km. The most downstream 4 km of this river channel is an open cross-section, formed with structural concrete, possessing a capacity of 270 m³/s, which corresponds to a return period of 20 years. Upstream of this section, the river is completely covered, to make space for a major traffic artery. Like in other similar cases, the decision to cover the Ilisos was taken immediately after the second world war, when economic constraints and rapid urban expansion pressed for quick solutions, ignoring flood risks as well as

J. Marsalek et al. (eds.), Advances in Urban Stormwater and Agricultural Runoff Source Controls, 141–146.
© 2001 *Kluwer Academic Publishers. Printed in the Netherlands.*

environmental and cultural considerations. Like the Kifisos, the Ilisos also transports high pollution loads to the sea.

2. Stormwater Drainage Network

The storm sewer protection network of Athens is about 1200 km long. It is estimated that the area needs a network of a total length of 3000 km. Natural streams flowing through the area have a total length of 120 km. Stormwater collectors are made of concrete, with a 40 cm minimum internal diameter for circular pipes and a 90 × 60 cm size for egg-shaped pipes. Common problems include obstructions and illicit connections of sewage.

3. Wastewater System

The wastewater sewer network of Athens extends over 3000 km. Sewage is treated in two treatment plants. The Metamorfosi's treatment plant treats biologically 8000 m^3/day of septic tank effluents and 13000 m^3/day of urban wastewater.

The Psitalia treatment plant is designed with a maximum capacity of 1 000 000 m^3/day of dry weather flow. It provides primary treatment, while the biological stage, plus phosphorus and nitrogen removal is under construction. A 2000 m outfall discharges treated effluents into the sea at a depth of 65 m. Dewatered sludge is disposed at a solid waste landfill.

4. Adaptation of the Wallingford Procedure to the Greek Conditions

The rapid expansion of Athens during the last four decades and the inadequacy of the stormwater drainage planning has led the Division of Water Resources of the National Technical University of Athens to examine the conditions for the adaptation and application of a model in Greek urban drainage networks. This could permit an optimal design of diameters and gradients of new pipe systems, analysis of the performance of existing systems, and the evaluation of alternative drainage systems. The model that was applied was the Wallingford procedure, initially developed for the United Kingdom [1]. This method of drainage design and analysis uses a mathematical modelling tool that can deal with combined and separate drainage systems, and models flow as well as water quality parameters.

For this purpose, the Argyroupolis pilot area, situated in the SE section of the greater Athens, has been selected and equipped with rain and flow recording gauges. The mathematical model has been set up taking into account the catchment characteristics and the data of the existing separate storm sewer network belonging to the Athens Water and Wastewater Company (EYDAP). The calibration of the model has been based on field measurements. Its verification has been tested by means of subsequent rain events. The above research has been realised in the frame of the European Sprint project, with the Greek participation covered by EYDAP.

An overall good agreement has been noted for simulated and recorded flow values, with respect to the peaks and the total volume of the storm events.

5. Detention Systems

As pointed out earlier, the constructed and operated storm sewer network extends to 1200 km, out of the 3000 km estimated as needed [2]. An immediate approach to the problem could be to proceed with planning the completion of the existing network, and construction of the 1800 km of missing collectors. However, this approach is unrealistic for the following reasons:

- Traffic conditions in Athens constitute a serious daily problem that can hardly tolerate additional obstructions from open excavation construction of sewage collectors.
- Extensive no-dig construction of collectors is beyond the budget capacity.
- Since the existing network and specifically the main collectors and natural streams consist of closed sections of limited capacity, exceptionally exceeding a return period of 10 years, an eventual completion of the secondary and tertiary part of the network would inevitably lead to the surcharging of the network and to serious flooding along the downstream flat area of the city.

Consequently, we have to consider:
- application of storage facilities, and
- source-control systems

5.1 THE ANO LIOSIA PROJECT

Ano Liosia is situated in the northwest periphery of Athens [3]. Before urbanisation, the area was drained by a network of seasonal streams that drained to a natural seasonal lake. Urbanisation took place in this area, ignoring natural streams and the receiving lake. As a result, severe flooding caused extensive property damage and life hazards. Recently, the municipality has proceeded to construct a small reservoir of 32000 m^3, mainly for recreational purposes.

Downstream of this reservoir, a main collector formed by a trapezoidal channel with an area of 16 m^2 and a capacity of 50 m^3/s connects Ano Liosia to the network of the greater area. The 50-year flow at this point amounts to 141 m^3/s.

In the frame of a research project, we considered the design of a detention storage facility that would be able to control downstream discharge up to a desired return period. For this purpose, the following tasks were done:

i.   i-t-T relations were established on the basis of local meteorological data.
ii.   Typical hyetographs were tested using the Huff's method (1, 2, 3 quartiles) and the method of alternative blocks.
iii. Hydrographs for the external catchments have been developed using the theory of the instantaneous unit hydrograph and the assumption of a linear reservoir (n = 1, 3).

iv.   Flow in the urban area has been simulated using the Hydroworks model of the Wallingford Procedure. Parameters of the model have been selected on the basis of the results of the Argirupolis pilot described earlier.

v.   Modelling served to route input flows and produce a final hydrograph at the entrance to the designed storage facility.

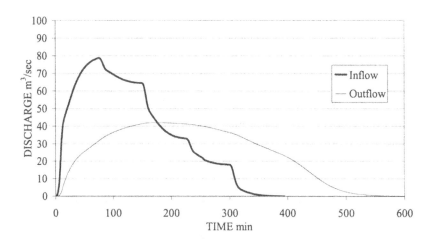

*Figure 1.*Final hydrograph and results of modelling inflows

vi.   Reservoir routing has been applied to estimate the storage volume corresponding to several outlet flows (T = 20, 50 years). The obtained storage volumes have been compared to the results obtained with a modified rational method, a Japanese method and the approximate method proposed by A. Paoleti.

For the total catchment, including 700 ha of non-urban and 300 ha of urban land, a storage facility of 500 000 m^3 has been found necessary to obtain an outflow discharge with a return period of T = 10 years for an inflow with T = 50 years. The above corresponds to a storage facility in the order of 500 m^3/ha of the whole catchment, or 125 m^2/ha of the urban catchment for an assumed depth of 4 m. The land cost in the study area amounts to 1 M Euro/ha and is on the low side of prices, which vary from 1 – 4 M Euro/ha for the city suburbs. On this basis, land acquisition for the construction of a storage facility at Ano Liosia imposes a total cost of 12.5 M Euro.

Including the construction expenses, the Ano Liosia installation will cost about 25 M Euro, or 83,300 Euro/ha of urban area, or 2,500 Euro/inhabitant.

5.2. SOURCE CONTROLS

The above presentation of the situation in the Greater Athens area leads to the conclusion that violation of the natural drainage network, construction of covered creek sections with high traffic overhead, and the extensive urbanisation serving the needs of more than 4 000 000 people, the detention storage facility needed represents an extremely high investment without guaranteed availability of the land needed.

Consequently, the possibility of introducing source controls for stormwater runoff should be seriously examined.

In the case of Athens, the factors favouring the implementation of source controls are the following:

- *Availability of land.* Urban planning regulations for town suburbs correspond to coefficients of coverage between 40 – 60%.
- *Soil permeability.* Geological formations pertaining to the area consist of altered schists and allow medium to high rates of permeability.
- *Vegetation.* Actual cultural and esthetic considerations favour the development of large grass-covered areas, enhancing the soil infiltration capacity.
- *Ground water level.* The year-round groundwater level is particularly deep, without significant seasonal variations.

Among the adverse factors, one can state the following:

- *Centralized stormwater management.* EYDAP, a central agency, is responsible for the stormwater management in the entire area. It is understandable that EYDAP is inclined to consider major central plant projects for water management, rather than individual decentralised source controls.
- *Public acceptance.* The public opinion already formed in accord with a centralised stormwater control agency is not prepared to consider individual fragmented initiatives.
- *Legal frame.* The legal frame to enforce applications of source controls on private property is lacking.
- *Technological experience.* Source controls represent a new technology in Greece.

Operational development of source control installations has the following prerequisites:

- Transfer of technology
- Research of local conditions
- Drafting specifications
- Standardisation
- Formation of specialised construction companies
- Standardisation of the maintenance practice
- Regulations for supervision and operation control.

In this context, the support from other countries that have already records of application and operation of source control systems constitutes a fundamental condition for the introduction of this technology in the country.

6. References

1. Aftias, E. and Tarnaras, E. (1998) Adaptation of the Wallingford Procedure to the Greek Conditions, *Scientific Publication of Technical Chronicles* (in press).
2. Aftias, E., Baltas, E., Tarnaras, E., Lazaridis, S. and Mimikou, M. (1999) Urban water management: *Country paper of Greece, Euraqua Scientific and Technical Review 6.*
3. Lazaridis S. (2000) Detention pond design for Flood protection of the Ano Liosia area, *Master Thesis*, National Technical University of Athens, Athens.

THE USE OF CONTINUOUS LONG TERM SIMULATIONS FOR THE DESIGN AND IMPACT ASSESSMENT OF SOURCE CONTROL MEASURES

G. VAES and J. BERLAMONT
Hydraulics Laboratory, University of Leuven
de Croylaan 2, B-3001 Heverlee, Belgium

1. Introduction

Since the new Flemish guidelines for urban drainage were introduced in 1996 [1], more emphasis has been put on source control measures in order to reduce the peak runoff from urban areas during wet weather conditions. The keyword is 'disconnecting' impervious areas from the combined sewer system. This involves the construction of upstream storage and infiltration facilities, rain-water tanks for reuse in households and the revaluation of ditches [2,3]. However, one must realise that the runoff discharges from the 'disconnected' areas still have to be taken into account, because during heavy rainfall periods not all rain-water can be stored locally. Therefore, urban runoff plans with respect to extreme events and flooding risks are very important.

This source control policy brings along different modelling requirements. The attitude changes from 'runoff transport as fast as possible' to 'runoff transport as slow as possible'. Because of the long emptying times of source control facilities, a long antecedent period of rainfall influences the design. Moreover, these facilities most often have an outflow which does not vary linearly with the storage. This non-linearity between the state and flow parameters leads to the situation where storage capacity is less dependent on the peak flows, but more on the complete variability of the rainfall input during a long antecedent period. Therefore, the frequency of the system outflow is not equal to the frequency of the system inflow. It is useless to predict the magnitude of effects (i.e. water level, discharge, etc.) when no accurate probability (i.e. frequency or return period) can be assigned to this effect. Because of the high variability of the rainfall, the required storage volumes can be assessed well only if continuous simulations with long rainfall series are performed.

2. Impact of Source Control

Urban stormwater runoff through combined sewer systems, as it is commonly applied in Flanders, creates large problems. The fast runoff through the pipe systems creates flooding problems downstream and leads to lower groundwater tables upstream. Furthermore, the use of combined sewer systems leads to diluted waste water at treatment plants and pollutant emissions into the receiving waters at combined sewer

J. Marsalek et al. (eds.), Advances in Urban Stormwater and Agricultural Runoff Source Controls, 147–157.
© 2001 *Kluwer Academic Publishers. Printed in the Netherlands.*

overflows. Source control measures are powerful tools and are able to tackle these problems if they are applied on a large scale. Source control measures not only have an impact on the low flows, but are also capable of reducing the frequency of extreme events. The disconnection from the combined sewer system of a relatively small part of the impervious area can lead to a significant reduction of the frequency of extreme events. On the other hand, an underestimation of the contributing area will lead to a significant underestimation of the return periods of extreme events (Figure 1) [4].

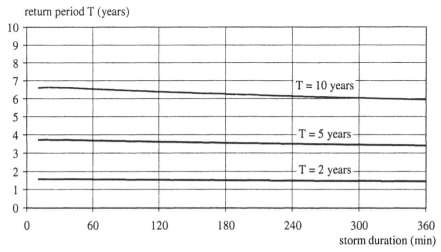

Figure 1. Reduction of return period in the case of 10 % underestimation of the water input (impervious area, rainfall intensity, losses, ...) into the combined sewer system (for design storms with return period of 2, 5 and 10 years).

3. Rain-water Storage Tanks

Using a simple reservoir model and performing continuous long term simulations, the optimal design volumes were determined for rain-water storage tanks as a function of the effective connected roof area and as a function of the reuse consumption in the household (Figure 2) [5,6,7]. The effective connected area is the horizontal roof area multiplied by a number of correction coefficients. Corrections are necessary to take into account the slope and orientation of the roof (Table 1), the losses on the roof as a function of the roof type (Table 2) and eventually the losses in the filter. Self-cleansing filters use a small part of the water to discharge the solids, which leads to reduction coefficients between 0.95 and 0.90 (5 to 10 % losses).

rain-water tank volume (m³/100 m²)

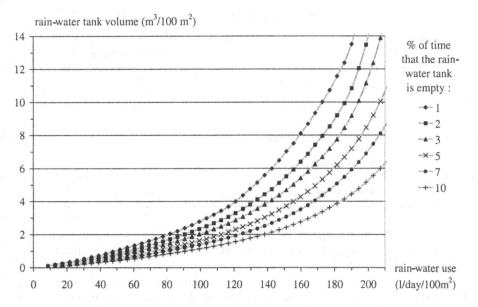

Figure 2. Design chart to determine the percentage of time that the rain-water tank is empty, the necessary volume of a rain-water tank or the possible rain-water use (all parameters are relative to the effective contributing area).

TABLE 1. Correction coefficients for Flanders on the effective contributing area as a function of the roof slope and orientation [8].

roof slope (degrees)	North-East	North-West	South-West	South-East
30	0.75	1	1.25	1
35	0.70	1	1.30	1
40	0.64	1	1.36	1
45	0.57	1	1.43	1
50	0.48	1	1.52	1
≥ 55	0.45	1	1.55	1

150

TABLE 2. Reduction coefficients for initial losses [9].

Roof type	Reduction coefficient
terrace roof with gravel	0.6
terrace roof with bitumen	0.7 to 0.8
sloped roof with slates and tiles	0.75 to 0.9
sloped roof with glazed slates	0.9 to 0.95
sloped roof with bitumen	0.8 to 0.95

The effect of storage in rain-water tanks on the combined sewer overflow emissions was investigated using reservoir models [5,6,7]. Figure 3 illustrates the conceptual model used. Figure 4 shows the effect of rain-water tanks on the necessary storage in the combined sewer system in order to obtain the same overflow frequency (the overflow frequency is the main criterion for the emissions at combined sewer overflows in Flanders). The required storage in rain-water tanks is about 7 times higher than in the case of in-line storage, but in-line storage is more expensive. Figure 5 shows that extra benefit can be obtained in flattening the peak overflow discharges when using rain-water tanks as compared with in-line storage in the combined sewer system, when the same overflow frequency is imposed.

Based on the positive effect of rain-water tanks that was found, the minimum storage capacity for the rain-water tanks is set to 5000 litres per 100 m^2 roof area and a minimum reuse connection of one toilet or one washing machine is imposed [2]. A smaller tank size could fulfil the reuse requirements, but the extra storage is imposed to maximise the retention. It is very important that the tank size is coupled to the size of the impervious area (and not a fixed tank size per house).

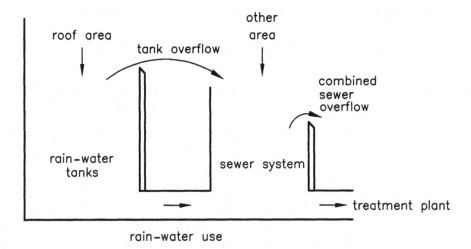

Figure 3. Schematic representation of the used reservoir model to assess the impact of rain-water tanks on the overflow emissions.

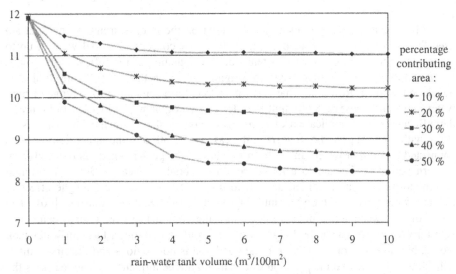

Figure 4. Required storage in the sewer system to obtain an overflow frequency of 7 days with overflow per year as a function of the rain-water tank volume (when a rain-water reuse of 150 litre/day/100m² is assumed).

peak overflow discharge (mm/h)

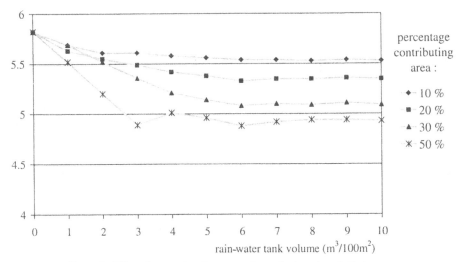

Figure 5. Effect of storage in rain-water tanks on the (mean) peak discharges
at the combined sewer overflow when the rain-water reuse is equal to 150 litre/day/100m²
(all combinations have an overflow frequency of 7 days with overflow per year).

Rain-water tanks can also have an effect on the more extreme events that are
used for the design of combined sewer systems. A conceptual model was set up to
incorporate the effect of rain-water tanks on the flattening of the inflow hydrographs
for the downstream drainage system (Figure 6). For this, a software tool was
developed (named 'Rewaput') that performs a long term simulation for a range of rain-
water tanks and converts the output into design hydrographs [10,11,12]. In Figure 7,
the influence of the application of rain-water tanks is shown on the Flemish design
storm with a return period of 5 years in the case where 30 % of the impervious areas is
connected to the rain-water tanks. In this example, the peak intensity is reduced from
a return period of 5 years to a return period of about 1.5 years. Both the original
design storms and the modified design storms, which take into account the effect of
the rain-water tanks, are based on intensity/duration/frequency relationships [10]. The
effect of rain-water tanks on the inflow into the combined sewer system cannot be
determined by using a single design storm and routing it through the rain-water tank
model, because the corresponding initial condition is not obvious and the first rainfall
will fill the tank, so that the peak intensities might not be influenced. The reason is the
non-linear behaviour of the rain-water tank system, which leads to a large influence of
the time variability of the rainfall. There is no straight relationship between input (i.e.
rainfall) and output (i.e. rain-water reuse consumption + tank overflow discharge).
Therefore, an accurate estimation of the effect can only be made based on long term
simulations with a statistical analysis of the tank outflow discharge.

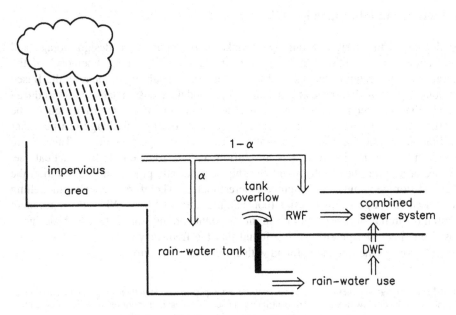

Figure 6. Conceptual model used to incorporate the effect of rain-water tanks
on the inflow hydrographs for combined sewer system design
(RWF = rain-water flow; DWF = dry weather flow).

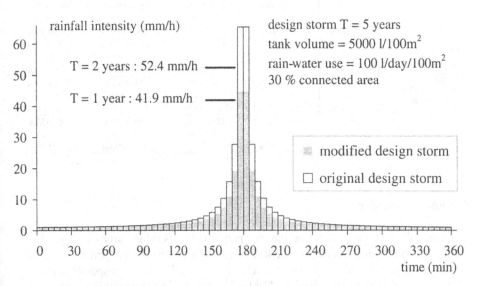

Figure 7. Effect of the installation of rain-water tanks on a large scale
on the design storm for a return period T of 5 years. This example shows
the effect if 30 % of the impervious areas is connected to rain-water tanks
with a volume of 5000 litres per 100 m² roof area and a rain-water use
of 100 litres per day and per 100 m² roof area.

4. Storage and Infiltration Facilities

Analogous to the design of rain-water tanks, it is important to design storage and infiltration facilities based on long term simulations. The long 'memory' of the storage facilities requires the use of the complete time variability of the rainfall. Based on long term simulations using a conceptual model, a design table was produced which gives the required storage volumes as a function of the return period of the overflow and the outflow discharge (Table 3). This outflow discharge can be either infiltration, throttled outflow, evaporation, etc. The storage volumes in Table 3 are based on the assumption of a constant outflow and are rounded off for practical use. In case of a more linear outflow relationship (e.g. throttle pipe with free outflow) the required storage volumes are slightly underestimated. Therefore, a specific modelling is advised for large storage basins (contributing area >10 ha). When the antecedent rainfall (or initial storage) is not taken into account and the storage facility is designed based on single design storms, it is found that the frequency of the overflow can be down to 50% of the frequency found with continuous simulation.

TABLE 3. Storage volumes (relative to the contributing impervious area; in three different dimensions) as a function of the outflow (relative to the contributing impervious area; in three different dimensions) and as a function of the return period of the emergency overflow.

		return period (overflow)			
		½ year	**1 year**	**2 years**	**5 years**
outflow : either by infiltration, evaporation and/or throttled throughflow	**10 l/s/ha** 3.6 mm/h 360 l/h/100m^2	**75 m3/ha** 7.5 mm 0.75 m^3/100m^2	**100 m3/ha** 10 mm 1 m^3/100m^2	**150 m3/ha** 15 mm 1.5 m^3/100m^2	**200 m3/ha** 20 mm 2 m^3/100m^2
	5 l/s/ha 1.8 mm/h 180 l/h/100m^2	**100 m3/ha** 10 mm 1 m^3/100m^2	**150 m3/ha** 15 mm 1.5 m^3/100m^2	**200 m3/ha** 20 mm 2 m^3/100m^2	**275 m3/ha** 27.5 mm 2.75 m^3/100m^2
	2 l/s/ha 0.72 mm/h 72 l/h/100m^2	**150 m3/ha** 15 mm 1.5 m^3/100m^2	**200 m3/ha** 20 mm 2 m^3/100m^2	**275 m3/ha** 27.5 mm 2.75 m^3/100m^2	**350 m3/ha** 35 mm 3.5 m^3/100m^2
	1 l/s/ha 0.36 mm/h 36 l/h/100m^2	**200 m3/ha** 20 mm 2 m^3/100m^2	**275 m3/ha** 27.5 mm 2.75 m^3/100m^2	**350 m3/ha** 35 mm 3.5 m^3/100m^2	**450 m3/ha** 45 mm 4.5 m^3/100m^2

5. Flow Attenuation

The Flemish guidelines prescribe that for new large impervious areas (starting from 0.1 ha), the downstream flow must be limited to 10 litres/s/ha for a return period which is a function of the sensitivity of the receiving water [1]. The corresponding storage volumes for this can be found in Table 3. For sensitive receiving waters this throughflow limitation can be further decreased.

However, limiting the throughflow to 10 l/s/ha for a small return period does not always lead to the optimal flow attenuation. It is more important to use a high return period for the overflow of the storage facility, than to limit the throughflow.

Figure 8 shows an example of the runoff to storage basins with a throughflow limits of 10 and 20 l/s/ha respectively with the same storage capacity. The throughflow limit of 10 l/s/ha gives a return period of 2 years for the overflow, while the throughflow limit of 20 l/s/ha gives a return period of 9 years. The latter case leads to a better attenuation of the peak flows, despite the larger throughflow limit. This shows that the throughflow limit only is not sufficient to optimise the attenuation of the downstream flow and that a case specific long term simulation can have a large benefit. For sensitive receiving waters, it is thus better to increase the return period of the emergency overflow than to further limit the throughflow.

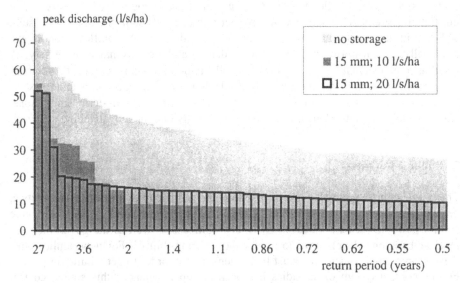

Figure 8. Distribution of peak discharges for the runoff of a catchment without a storage facility as compared with two scenarios with storage facilities (both 15 mm of storage and respectively 10 and 20 l/s/ha throughflow)

6. Design of Ditches

The revaluation of ditches is one of the main topics in the source control policy in Flanders [1,3]. The main function of the ditches shifts from runoff to storage and infiltration, but the runoff function at peak flows must be assured without flooding. Therefore, for the design of ditches, a two step approach is proposed [3]. First, the storage volume in the ditch must be large enough to store and infiltrate the water from locally connected areas. This means that they have to be designed using the design table (Table 3). In the second step, a hydrodynamic control simulation is performed for the rain-water system. The hydrodynamic simulation is necessary in order to take into account the backwater effect and the interaction with pipe systems. For this control simulation, no flooding may occur for a return period of 5 years. The infiltration flow is neglected as compared with the peak flow in the drainage system and the initial conditions must be chosen according to the outflow conditions. Eventually, design storms with longer antecedent rainfall periods may be necessary. A simultaneous evaluation of the storage and runoff function is very difficult or practically impossible with the currently available modelling tools and is unnecessary. Moreover, the storage function must be evaluated based on long-term simulations, which would require too long calculation times in a hydrodynamic modelling system.

7. Public Awareness

The source control policy looks nice on paper, but is difficult to implement in practice, because many people are involved. One of the main principles is that the source control policy must be applied on all levels. Citizens as well as the government on local and regional levels have to accept their responsibility. For new applications, these source control measures must be implemented in order to get a building permit. Furthermore, a system of subsidies has been set up to enhance this source control policy in other cases. In order to make the people and local authorities aware of the needs and possibilities and to give them practical information, brochures are prepared. Specific brochures are made for each group of people, e.g. local communities, architects, citizens, etc. and contain tips for rational water use as well as technical information on how people can implement this in practice [13,14,15]. Furthermore, workshops are organised for engineers and decision makers to inform them about the new technologies and methodologies and a task group on sewer system concepts exists already for years within the Flemish sewer system and waste water treatment forum 'Vlario'.

8. Conclusions

Source control measures are an ideal tool to solve problems with excessively high peak flows from urban catchments. They not only reduce the emissions at combined sewer overflows, but also have a significant impact on extreme events and the dimensions of the combined sewer system. The long antecedent period that influences storage facilities requires the use of the full time variability of the rainfall in order to obtain an accurate design. The use of design storms will lead to an underestimation of the storage volumes. For the Flemish applications, design rules have been successfully established based on continuous long-term simulations using conceptual models.

9. References

1. Anonymous (1996) *Guidelines for an integrated sewer system management in Flanders, code of good practice* (in Dutch), VMM.
2. Anonymous (1999) *Guidelines for an integrated sewer system management in Flanders, code of good practice for rain-water tanks and infiltration facilities* (in Dutch), VMM.
3. Anonymous (1999) *Guidelines for an integrated sewer system management in Flanders, code of good practice for revaluation of ditches* (in Dutch), VMM.
4. Vaes, G. and Berlamont, J. (1999) The effect of changing technology on combined sewer system design, *8th International Conference on Urban Storm Drainage*, Sydney.
5. Vaes, G. and Berlamont, J. (1998) The effect of storage in rain-water tanks (in Dutch), Hydraulics Laboratory, University of Leuven.
6. Vaes, G. and Berlamont, J. (1998) Optimization of the reuse of rain-water, *International WIMEK Congress on Options for Closed Water Systems*, Wageningen.
7. Vaes, G. and Berlamont, J. (1999) The impact of rain-water reuse on CSO emissions, *Water Science & Technology* **39 (4)**.
8. Simons, W. (1995) Rain-water tanks (in Dutch), *Beton* **129**, FeBe.
9. Van den Bossche, P. (1997) *Rain-water use* (in Dutch), VIBE.
10. Vaes, G. (1999) The influence of rainfall and model simplification on the design of combined sewer systems, PhD thesis, University of Leuven.
11. Vaes, G. and Berlamont, J. (1998) The effect of storage in rain-water tanks, part 2 : the effect on the design of combined sewer systems (in Dutch), Hydraulics Laboratory, University of Leuven.
12. Vaes, G. and Berlamont, J. (2000) The effect of rain-water storage tanks on design storms, 1st international conference on urban drainage on internet.
13. Vaes, G. and Van de Veire, P. (1998) *Reuse of rain-water* (in Dutch), Vlario.
14. Michielsen, K., Vaes, G., Vancalbergh, L., Versweyveld, G., Demuynck, W., and Dupont, G. (1999), *Disconnection of impervious area, storage and infiltration* (in Dutch), Vlario.
15. Vaes, G., Creffier, W., Vlerick, Ch., De Backer, L. and Van Peteghem, M. (2000) *Water guidelines for architects* (in Dutch), VMM.

THE ROLE OF STORMWATER SOURCE CONTROL IN THE CONVERSION OF THE NATIONAL HOSPITAL OF NORWAY INTO AN ECOLOGICAL RESIDENTIAL COMPLEX

T. ANDERSEN[1], J. EKLUND[2], T. TOLLEFSEN[3], T. LINDHEIM[4] & W. SCHILLING[5]
[1]STATKRAFT GRØNER, P.O. Box 400, N1327 Lysaker, Norway
[2]Grøner AS, Oslo, Norway
[3]Statsbygg, Oslo, Norway
[4]Bjørbekk & Lindheim Landscape Architects, Oslo, Norway
[5]Norwegian University of Science and Technology, Trondheim, Norway

1. Project Description (Pilestredet Park)

The National Hospital of Norway ("Rikshospitalet") moved to a new building complex on the outskirts of Oslo in spring 2000. The old hospital is situated in the very centre of Oslo and covers an area of about 7 ha. Its buildings were built over the past century and comprise over 110,000 m^2 of floor space. About half of the complex is to be demolished, and there will be about 85,000 m^2 of new construction while 50,000 m^2 will be renovated. About 60 % of the new or reconstructed area will be residential, while the rest will consist of offices, schools, or commercial premises of different types. The outdoor areas will be landscaped as public spaces.

It was decided in 1997 to use this area (named "Pilestredet Park") as a pilot project for urban ecologic development [1] [2]. The project is done in collaboration with Statsbygg, the national building authority, and the municipality of Oslo as the main partners. Collaboration of different project teams with a wide range of professional backgrounds is anticipated.

The main goals of the Urban Ecology Program were:

- Improve urban environmental quality, including green spaces, reduction of traffic, reduction of noise, etc.
- Promote infrastructure for public transport and pedestrians, district heating, and waste separation.
- Apply selective demolition including sorting, recycling and minimum transport of demolition wastes.
- Re-use existing buildings to a maximum extent, including specific projects where re-used materials are applied.
- Design outdoor areas in an integrated way as an urban ecology park.
- Apply environmentally-friendly architecture including energy design, healthy materials, natural air-conditioning, etc.
- Conserve energy, including renewable energy, and impose strict consumption limits.

159

J. Marsalek et al. (eds.), Advances in Urban Stormwater and Agricultural Runoff Source Controls, 159–168.
© 2001 Kluwer Academic Publishers. Printed in the Netherlands.

The treatment of stormwater was a part of this urban ecology project, and items mentioned above as, for example, landscape architecture, re-use of materials, and infrastructure in general were of interest in developing solutions concerning stormwater management.

2. General Information about the Planning of Outdoor Areas

Statkraft Grøner, the consulting engineers, together with Bjørbekk & Lindheim, the landscape architects, have been responsible for the planning of outdoor areas, i.e., the landscape architecture, roads, pedestrian walks, etc., and the system for stormwater collection and treatment. Additionally, the waste separation and collection systems were included in our contract.

The planning of the outdoor areas was carried out in three main phases:
- Preliminary studies (August - September 1999)
- Pre-feasibility studies (October - November 1999)
- Detailed planning (started in November 2000).

Traditionally, there has been some conservatism among the municipal experts and consulting engineers working in the field of urban water management, applying non-conventional techniques. However, pilot projects, including the one described here, demand that project participants must be open-minded with respect to new solutions.

In the project team for the outdoor areas, the landscape architects were given considerable freedom in the first stages of the project, and the engineers then tried to find technically feasible solutions to solve possible problems. Examples of this approach include infiltration, roof gardens, above-ground storage tanks, under-ground storage tanks (utilisation of existing structures), water falls, open basins, open channels, natural trenches, reuse of stormwater for washing, toilet flushing and irrigation, all with the aim of reducing runoff and increasing retention time of stormwater, before its discharge to the municipal combined sewer system in the surrounding streets.

The outdoor areas have been planned as an integrated landscaping project, applying ecological principles. Most of the outdoor area will be public. It was considered important to create a variety of open spaces, ranging from fully public areas to semi-private courts in the residential quarters, including small private gardens for some of the ground floor apartments. Also in its appearance, the landscape should express variety ranging from wilderness (including wetlands) to garden-type areas.

Green roofs (with roof-gardens) were also included in the plan, motivated by the general perception that a greater extent of green areas improves the air quality and local micro-climate in general.

The landscaping project aims, in imaginative ways, to re-use old building materials: stones, timber, bricks, etc. for pedestrian walks, seating arrangements, walls, outdoor pergolas and furniture. In this way the history of the area would be preserved and visible in the new landscape.

Water becomes another central feature of the outdoor areas; its cycle will be expressed clearly, and its different qualities will be used, again in an imaginative way, to show that it is a resource rather than waste! For example, some rainwater will be stored for watering the outdoor areas. The open waterways with waterfalls and a wide variety of designs will be used as a distinctive expression in the landscape design. So far, the main principles of the landscape design were described. Stormwater source control will be emphasised below and described in more detail.

3. Requirements for Stormwater Management

Regarding stormwater runoff, the local authorities (the Municipality of Oslo) have the intention to maintain a natural water balance, and have defined the project goal:
- The maximum stormwater flow shall be less than 10% of the maximum flow from an urban area with a traditional drainage system.

Due to legal restrictions, this requirement could only be defined for the new buildings and areas, but we tried to achieve this goal also with respect to the existing buildings, whenever possible. These requirements were also included in the contract documents, when Statsbygg assigned different parts of the redevelopment area to different contractors.

Given the catchment parameters, the concentration time was found to be 7 minutes. Thus, a design storm of 7 minutes duration was chosen to estimate the maximum expected runoff rate for a given return period.

Intensity-duration-frequency (IDF) curves from the Norwegian Meteorological Institute were used, for the Blindern station located just a few kilometres north of the area. These statistics include return periods from 2 to 100 years, and rainfall durations ranging from 5 to 360 minutes. In Oslo it is a common practice to use a return period of 10 years to size storm sewers, while the European standard EN 752 requires a return period of 2 years for small residential areas (CEN, 1993 [3]).

The maximum flow rates of runoff from the project catchment were calculated as shown below:

TABLE 1 Calculated maximum flow

Rainfall duration	Rainfall intensity (10-year return interval)	Calculated maximum flow (l/s)
5 minutes	322 l/s ha	1160
10 minutes	224 l/s ha	810
7 minutes	**280 l/s ha**	**1000**

162

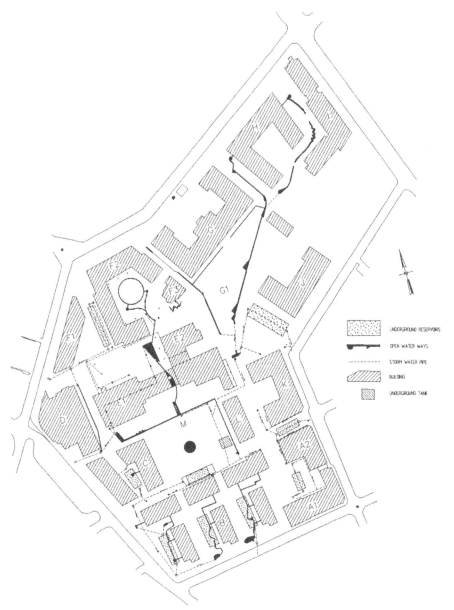

Figure 1 *Plan view of the Pilestredet Park project area (total area = 7 ha)*

    The requirement that the maximum runoff rate should be reduced by 90% (to 10%) of the maximum from an urban area with traditional drainage systems meant that the peak runoff for the whole catchment must not exceed 100 l/s.

Since the area is divided into several subcatchments, where different contractors are involved, the permitted runoff for each of those was set to 100 l/s multiplied by the corresponding percentage of the total area.

As retention is one of the available measures to reduce the peak runoff rate, the applicable time of concentration would increase and, thus, a longer duration rainfall could become the critical loading input.

4.   Applicable Methods for Source Control

The main principle of source control was found to be retention of stormwater runoff in different forms. As discussed above, infiltration was not a viable option, because of the low perviousness of soils in the project area.

Another feature was increased evapotranspiration by trees and other plants, which was further enhanced by the main goals that were defined regarding the landscape design (i.e., more plants than usual, open waterways with waterfalls, etc.). Also the green roofs will have an important effect with respect to runoff reduction.

The retention measures investigated are discussed below.

4.1   UNDERGROUND RESERVOIRS

Such reservoirs can be placed in parks, under roads or under open areas. The reservoirs have to be deeper than the freezing zone (winters in Oslo are cold, with repeated thaw cycles). The sides and the bottoms of the reservoirs are made waterproof by using a layer of clay, in order to avoid the problems described above. The reservoir bodies are made of crushed stone and rock such that the effective storage consists of the pore volume between the rocks. The effective storage was estimated to be 20-25% of the total volume of the reservoir.

The stormwater from buildings and landscape is first conveyed through sand traps to an inlet basin. From this basin the water enters the reservoir through two or three perforated pipes. At the downstream end of the reservoir, similar pipes, about one metre lower than the inlet pipes, drain the reservoir into an outlet basin for final drainage out of the project area. The basin at the outlet of the reservoir will be used to measure the outflow rate. Should this rate exceed the permitted limit, a vortex valve will be installed to throttle the flow, if necessary.

As an alternative to the crushed stone fill, and to increase the re-use of old materials, it was proposed to use also crushed concrete and/or crushed bricks in the reservoirs [4] [5]. Crushed concrete could be used only if it were washed after crushing. Stormwater runoff in Norway is often acid due to the low buffer capacity of soils. Experience from some practical tests shows that there would be chemical elutriation from the concrete, and that pH would increase at the outlet of such reservoirs up to about 9.0. Also, not all the cement in the concrete may be chemically bound. Experience from some test roads shows that free cement in recycled concrete tends to be activated later. Thus, there is a possibility that the reservoir may gradually become clogged and its porosity be reduced.

Bricks can absorb water, which yields an additional positive retention effect. However, the structure of crushed bricks is "needle shaped", and there is a tendency that it further disintegrates during operation, especially in winter conditions. Despite these concerns, the re-use of concrete and bricks will probably be applied in full scale in some of the retention reservoirs, in order to gain more experience with this technology.

4.2 RETENTION ON FLAT ROOFS

Stormwater retention on flat roofs is of interest where other retention methods are difficult to apply, for instance, if the area around the buildings does not allow to install underground reservoirs.

Obviously, it is essential that the watertight layer is of good quality, and that it is well protected against damage. This protection could be provided by a 10-cm layer of clinker, crushed bricks or crushed concrete that would also serve as intermediate storage. The mechanical strength of the material is not so important here. However, concrete was considered to be a little more risky due to chemical elutriation by acid rain, and potential effects of increased sediment fluxes on downstream pipes. Crushed brick and clinker were considered to have about the same properties, but to increase the re-use of materials, our final choice was crushed bricks. The rainwater that will not disappear by evaporation, absorption or consumption by plants will be collected and transported away by downpipes.

Above the layer of crushed brick, we have recommended to place a separation layer of fiber-cloth and a top layer of soil to allow for plant growth and create green roofs.

It was calculated that a reduction of the maximum flow rate by 90% could be achieved if 60-80% of the roof area of the complex would be designed in such a way.

4.3 TANKS (UNDERGROUND OR INSIDE BUILDINGS)

In addition to the above measures, one central underground tank will be also provided. Flow from open channels/ditches from the eastern part of the area will discharge to this tank, thus providing retention storage for most of the eastern half of the area. This was considered necessary, because many of the existing buildings in this area will be retained. As they have sloping roofs, rooftop storage is impossible, and other control methods were considered difficult to apply.

The retained stormwater in the tank will be available for irrigation of the public park area nearby.

A control system will be installed to operate the tank for both retention (emptying after a rain) and irrigation. The tank volume is approximately 600 m^3, of which 200 m^3 are reserved for retention and 400 m^3 for irrigation. The retention volume is equivalent to approximately 9 mm of storage, related to the upstream impervious area, and its retention volume is equivalent to 100% of the design runoff volume of the connected area.

4.4 WATER TOWERS

The idea to install water towers was put forward by the landscape architect. The main motivation was to combine roof runoff retention and water storage to allow for extended flow periods in the waterways during dry weather. Water release shall be controlled such that, for example, every half an hour, flow would be generated for a duration of a few minutes. As only clean water shall be stored, no roof gardens were connected to the water towers.

The water towers have an inner diameter of about 3 m, and a height of about 10 m. The water is collected from a roof and led into a water tower situated at the corner of the building. There is a free drop of a few metres as the building is 2-3 m higher than the top of the water tower.

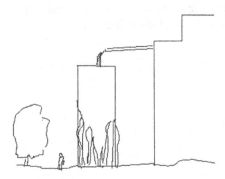

Figure 2   Principle sketch of a water tower

About 80 % of the volume is to be used for storage and successive discharge into the waterways. The remaining 20% is kept empty to have spare volume available during extraordinary runoff events.

In winter conditions, the outlet of the tower would be held open at all times to avoid freezing problems. It was also regarded as necessary to use thermostatic electric heating near the outlet area to prevent freezing of the outlet. In non-winter conditions, the outlets of the tanks will be controlled by electrically operated valves. The water towers were designed to be located at the sites F1 and H (Fig. 1).

4.5 OPEN WATERWAYS

From the beginning of the planning process, water was considered to be a positive element of the landscape, and the landscape architect applied water as a distinctive feature of the outdoor design. Two open waterways (east and west) pass through the area, both sloping southwards (see Fig. 1). The waterways have different types of chutes and waterfalls and feature some small ponds.

Water from the eastern waterway is discharged into the underground tank described above.

The open waterways would tend to increase the evaporation, and also have some small retention effect (lower velocity compared to pipes, some extra storage volume in the ponds).

5. Implemented Solutions and Current Experience

Since the drainage conditions in the area are complex and vary a lot between the different sub-catchments, we had to use a combination of drainage methods as described above. Space and other constraints restrict the free application of measures, e.g., using underground reservoirs depends on the availability of a sufficiently large open area. Where sufficient space was not available, e.g., close to some of the existing buildings, water retention tanks inside the buildings were regarded as the only solution required to achieve the 90% peak flow reduction.

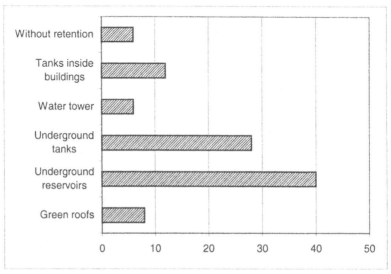

Figure 3   Distribution of different retention methods used (as percentage of total areas). Waterways are not included.

The history of the project so far has been that the preliminary studies were followed by two "intermission phases", and by the pre-feasibility studies. The first intermission phase was an environmental follow-up program for the whole area, and the second one was dedicated to formulate requirements and specifications for the outdoor design and the stormwater control for three of the sub-catchments, which were sold to contractors in late 1999 (B, C and H in Fig. 1).

After finishing the pre-feasibility studies in November 1999, the municipality and Statsbygg evaluated the project. There were some concerns about possible problems concerning freezing of the open waterways in winter conditions, but referring to the experience in Sweden and USA [6] [7] [8] [9] [10] [11] [12], this was not considered to be a problem. Even if some non-traditional solutions were used, the concluding statement from the municipality was rather positive.

The pre-feasibility study was approved in February 2000. As the municipality will be responsible for most of the operation and maintenance (O&M) of the outdoor area, there are still ongoing discussions between Statsbygg and the municipality about the consequences of this novel design with respect to O&M, so the detailed planning was halted until November 2000.

The contractors that will develop the first sub-catchments have just started the design process, and being the project team for the outdoor works, we have been involved in ensuring cooperation and approving the final design.

The main problem so far is that some of the features and conditions of the pre-feasibility study were not fully included in the sales conditions and the associated contract. The water towers will probably be difficult to implement due to this problem. Other problems may be caused by the need for more underground parking. Thus, there will be less room for underground stormwater reservoirs. The central underground tank will become smaller, due to technical difficulties to bring water from the eastern waterway all the way down to the tank.

Hence, the project will be somewhat reduced compared to the original intentions formulated in the pre-feasibility study. This is mainly related to unfavourable physical conditions and due to the relaxation of some of the strict initial requirements.

6. Results and Discussion

Our study shows that it may be possible to reduce the stormwater runoff by up to 90% from a heavily urbanised inner city development area by using different source control methods. The applicability of some of the methods depends on the availability of some open space or outdoor areas, while the use of indoor tanks, water towers or green roofs may be used under any circumstances.

Since the rebuilding of the area has not yet started, we have no operating experience from the use of these different methods. As underground reservoirs, underground tanks, and green roofs have been used in different other projects, there is a belief that the experience will be positive.

Even if we prefer simple systems for the sake of simplified operation, some of the methods have to rely on more sophisticated control systems (i.e., water towers or retention tanks).

7. Further Work

In the forthcoming detailed planning phase for the public areas and in the ongoing co-operation with the contractors for different sub-catchments, we will try to apply the proposed methods for stormwater control and the re-use of building materials as close to our intentions as possible.

It may be difficult to fully evaluate the effects of the different methods, especially to confirm that we have indeed achieved the 90% peak flow reduction. However, it is planned to install flow measurement equipment, relate the measured flows to rainfall intensities, and compare conventional drainage versus source control runoff scenarios.

A very important byproduct of the planning process was the fact that it opened our minds with respect to the feasibility of unconventional source control methods applied to inner city drainage. Today we are convinced that this approach is suitable for application in other similar projects.

168

8. References

1. Statsbygg og Oslo kommune (1998) Fra sykehus til sunne hus, Byøkologisk program for Pilestredet Park.
2. Statsbygg/ Oslo kommune, The Pilestredet Park urban ecology project, Oslo, Norway (paper).
3. CEN (1993) Drain and Sewer Systems Outside Buildings - Part 4: Hydraulic Design and Environmental Considerations, European Standard, EN 752-4.
4. Lahus, O., Jacobsen, S., and Myhre, O. (2000) Bruk av resirkulert tilslag i bygg og anlegg – status 2000 RESIBA-prosjektrapp. 01/2000, Pub.ID: 000074, prosjektrapp. 287.
5. Mastrup, A.-L. and Sundbye, T.-A. Resirkulert tilslag som fundament og omfylling i VA-grøfter, RESIBA-prosjektrapport.
6. Bäckström, M. (1999) Personal communication.
7. Bäckström, M. and Viklander, M. (2000) Integrated stormwater management in cold climates, Journal of Environmental Science and Health.
8. Bäckström, M. (1999) Porous pavement in a cold climate. Licentiate thesis 1999:21, Luleå University of Technology, Luleå, Sweden.
9. Stenmark, C. (1991). The function of a percolation basin in a cold climate. Proceedings of the 5th Int. Conf. In Urban Storm Drainage, Suita, Osaka, Japan, July 1990, pp 809-814.
10. Stenmark, C. (1992). Local disposal of storm water in cold climate. Licentiate Thesis 1992:24, Luleå University of Technology, ISSN 0280-8242, Luleå, Sweden.
11. Jansson, E., Lind, B., and Malbert, B. (1992) Lokal dagvattenhantering. Erfarenheter från några anläggningar i drift, VA-Forsk, Rapport 1992-09, VAV, Stockholm.
12. Sveinung, S., Milina, J., and Thorolfsson, S.T. (eds.)(2000) Urban Drainage in Cold Climate, report by UNESCO.

COPPER REMOVAL IN INFILTRATION FACILITIES FOR STORMWATER RUNOFF

M. STEINER and M. BOLLER
Swiss Federal Institute for Environmental Science and Technology,
Überlandstrasse 133 CH - 8600 Dübendorf, Switzerland

1. Introduction

Copper, zinc, lead and cadmium in roof and road runoff can be considered as major sources of heavy metals in urban drainage systems. Depending on the drainage system, they cause different environmental problems. In a combined sewer system heavy metals reach the wastewater treatment plant (WWTP) or can be discharged directly into a river as consequence of intense rainfall events. In the WWTP, heavy metals are transferred into the sewage sludge, to the receiving waters and into aquatic sediments [1]. If the sludge is used for land spreading in agriculture, heavy metals are distributed diffusely in agricultural soils where they accumulate slowly and may affect soil fertility [2].

Depending on the chemical and physical properties of heavy metals, up to 30% of the inlet load of a sewage treatment plant can reach the receiving waters [3]. Also, through this pathway, heavy metals are distributed diffusely in sediments and downstream river flow. In receiving waters, heavy metals can be toxic to algae and fish, both at very low concentration levels. It can be clearly seen that in both pathways, transfer into sewage sludge or receiving waters, we deal with very diffuse slow and long-term accumulation processes leading to problems in a time scale of 100 years and more. Therefore, the accumulation problems involved are not well recognised. According to the new (1992) Swiss water protection law, surface runoff from both roofs and roads should be infiltrated if possible. As a consequence, the heavy metal fluxes are redirected to infiltration sites. Their fate depends on the construction of these sites and may cause new types of relatively fast metal accumulation problems.

2. Infiltration Techniques

Generally, two infiltration techniques can be distinguished [4]. On one hand, the stormwater runoff may be directed to sub-surface infiltration systems in which the topsoil is omitted as infiltration layer. This technique distributes heavy metals diffusely in the ground in poorly defined layers 2-3 m below the ground level. In

169

J. Marsalek et al. (eds.), Advances in Urban Stormwater and Agricultural Runoff Source Controls, 169–180.
© 2001 *Kluwer Academic Publishers. Printed in the Netherlands.*

addition, heavy metals may reach the groundwater because the retention capacity of subsoils is generally lower than in topsoils.

On the other hand, infiltration may take place directly in topsoils. Because of a pronounced adsorption capability of these humus containing soil layers, heavy metals may be retained to a large extent, if the pore system of the soil layer allows for a relatively homogeneous water flow pattern [5]. In reality, the retention capacity can be seriously affected by preferential flow paths. Results of an EAWAG research project revealed that in a newly constructed infiltration pit, which was equipped with sampling ports along the depth, considerable heavy metal transport into deeper soil layers was observed. Although the topsoil layer was built in carefully with up to date technology, very fast breakthrough of copper and other substances could be measured during rain events. Another disadvantage of using natural soil as water purifying layer is its contamination. If heavy metals reach a certain level in topsoils, they will affect plant fertility, an indispensable natural resource. It can be concluded that none of the mentioned infiltration techniques meets the goals of intermediate and long term sustainability [6].

3. Source Control as Solution

A sustainable path can be achieved only if the use of materials with problematic effects on the environment can be avoided in future. Such changes in the use of construction materials involve various decision-makers in industry, government and business and their implementation will take decades. Until this goal is reached, the impact of heavy metals on the environment will continue and should, therefore, be minimised. It should be possible to control the spatial and temporal distribution of hazardous substances. One feasible way is to control the heavy metal fluxes at the infiltration sites. The retention of heavy metals should be managed with a technology as simple as possible. As a new element in urban drainage systems, we suggest to introduce a special adsorption layer at infiltration sites with an especially high affinity for heavy metals in all kinds of surface runoff. The artificial granular layer should have a sufficient hydraulic conductivity and should allow optimum efficiency and lifetime of the adsorption media and minimise hazardous waste production. Further, after exhaustion of the adsorption capacity, the adsorbent should be easily removable and its regeneration should be possible. The use of soil as adsorbent should be excluded and the contamination of neighbouring soils has to be avoided. Such adsorption layers may be a part of new types of infiltration facilities.

4. Copper Concentrations and Loads in Roof and Façade Runoff

In general, copper concentrations in roof runoff correlates with the area of copper plates used on a roof. The average concentration in roof water from a large area is considered to be about 90 µg Cu/l in average. This value corresponds to a copper covered area of 5% of the total roof area. This percentage is typical for roofs with a copper gutter. If a roof is covered completely with copper plates, average concentrations of 1800 µg Cu/l and more are usually observed. However, runoff events usually show marked first - flush effects characterised by up to one order of magnitude higher concentrations during the first 2 mm of runoff [7]. Figure 1 shows examples of typical copper concentration patterns in the runoff from a roof without copper installations, from a roof with copper gutters, and from a copper-plated roof.

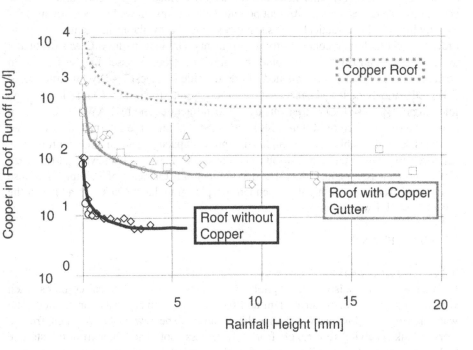

Figure 1. Copper concentration in different roof runoffs.

Copper is widely used as a plumbing material in exterior parts of buildings. Nowadays, copper loads from roofs account for up to 40-50% of the total copper load of combined sewer systems [3]. Discussions with decision-makers on the use of other materials are in progress and in some professional groups of architects and plumbers rethinking takes place. Although alternative materials may be available at the present time, it will take more than 50 years to replace larger parts of copper installations. Therefore, the application of a technical end-of-pipe solution is promoted to control the diffuse heavy metal fluxes in stormwater runoff. At the same time, additional efforts to retain heavy metals at infiltration sites may hopefully provoke rethinking in the building industry and promote the use of alternative materials.

5. Adsorbtion Materials

5.1. INTRODUCTION

A literature review on possible sorbents with especially high adsorption capacities in soils revealed three classes of materials: zeolites, humic acids and metal (hydr)-oxides such as those of aluminum and iron [8, 9, 10, 11, 12, 13]. Further, many experiments have been carried out with activated carbon [14]. Data from other experiments with cellulosic materials and other materials are also described [15]. A comparison of the adsorption capacities among different potential adsorbents shows best performance for iron hydroxide. For practical use, substances must be available in granular form in order to fulfil hydraulic conductivity requirements. Granular hydroxide was evaluated as a potential adsorbent and could be obtained from Wasserchemie GmbH in Osnabrück. Chemically, the obtained iron-hydroxide has mostly a crystal structure of β-FeOOH and is called akaganeite [16]. The crystal structure accounts for a large internal porosity. BET measurements by the manufacturer and by EAWAG resulted in a high internal surface of 300 m^2/g. It is assumed that the large internal surface of the particles is responsible for the high adsorption capacity. The marketed product is supposed to have grain size distribution from 0.3 to 2.3 mm. However, particle size distribution measurements suggest a maximum grain diameter of 0.8 mm. Because adsorption processes are highly dependent on pH, granulated $CaCO_3$ is added to the FeOOH in a 1:1 weight ratio.

5.2. PILOT PLANT

5.2.1. Construction

In order to test the adsorbents, a pilot plant consisting of 7 identical columns, each with a length and a diameter of 1 m and 0.05 m, respectively, was constructed. The water is directed downward through the columns. The raw water is taken from a storage tank receiving roof runoff from a fibreglass roof. The water from the storage tank is pumped into a small (12 l) trough-flow vessel 1 m above the columns, from where the supernatant flows back to the storage tank. All columns are fed from the trough-flow vessel. The flow rate in each column can be adjusted and samples can be collected of the effluent of the column as well as from several profile sample ports along the column. To avoid algae growth, storage, tubes and columns are covered to avoid any light induced growth.

5.2.2. Inlet

Roof water was sampled from a roof with an area of 100 m^2 and a copper gutter. According to runoff measurements, the average concentration reached about 300 µg Cu/l. Depending on the column feed, sampled roof water has to be pumped every week from the main storage tank (12 m^3) into the intermediate storage tank (0.6 m^3) from which the columns are supplied. In order to guarantee constant inlet conditions, the main storage was filled in autumn 1999 and from then on, the roof inflow could be disconnected. In June 2000, the storage had to be refilled. At this point in time, a shift to lower pH could be observed. The copper concentration in the main storage tank ranges from 20 to 30 µg Cu/l. In order to operate the columns with a concentration of 300 µg Cu/l, copper in the form of $CuSO_4(H_2O)_5$ was added. Filtration (0.45 µm) and subsequent analysis of the inflow shows, that at least 85 % of the copper is found in the filtrate, although considerable amount of particles are washed off from the roof. Chemical equilibrium calculations suggest that within the observed pH (7.1 - 7.7) in solution, Cu solubility is not critical.

5.2.3. Results

Different experiments have been carried out. Basically, constant and variable flow conditions should be distinguished. The first set of columns has been operated with constant flow in order to establish breakthrough curves. Four columns are used and filled with different adsorbents and one is left empty as a blank. In order to quantify and compare the performance of the $FeOOH/CaCO_3$ mixture, one column was filled with $CaCO_3$ granules, one with quartz sand and one with pure $FeOOH$ granules. The media bed height was 0.30 m in each column and the flow was set between 0.2 – 0.4 l/h. The experiments started in May 1999. Since then the columns are under steady flow. However, certain flow fluctuations due to temperature effects occur but do not affect the experiment. Copper effluent concentrations of columns 1 to 4 in the course of time can be seen in Figure 2.

Inflow and Effluent Copper Concentration of Columns

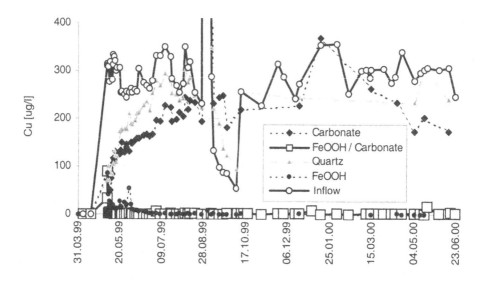

Figure 2. Inflow and effluent concentrations of columns 1 – 4.

The performance of the $CaCO_3$ media and the quartz sand column is not very good. The overall adsorption performance in these columns reaches 39% and 26%, respectively. It becomes clear that the columns containing FeOOH perform the best. For the mixed media column, the average effluent concentration was between 0.5 - 2 µg Cu/l. A removal efficiency of 99.3% can be calculated for the period from May 1999 until June 2000. The outstanding performance was attained during a loading period comparable to a rainfall amount of a 20-year return period into an infiltration shaft with a roof to infiltration area ratio of 100:1. All effluent samples contained only copper in the dissolved form. The FeOOH column performed the second best. An overall removal efficiency of 96% was achieved. Due to hydraulic problems, the loading decreased constantly and the column had to be taken out of operation in June 2000. Compared to the FeOOH/$CaCO_3$ column, adsorption performance of the pure FeOOH media was inferior especially at the beginning of the experiment.

The mixed media column was able to stabilise the pH whereas the FeOOH column showed a marked pH decrease. It is suggested that the pH decrease is caused by selective adsorption of CO_3^{2-}. However, it cannot be said clearly if the lower adsorption performance in the FeOOH column is caused by the lower pH. On the other hand, one can argue that even with a pH as low as 4, adsorption efficiency remains about 96%. This is an interesting and favourable observation, because in this pH range adsorption usually does not occur onto other solid iron-hydroxides such as FeOOH [17, 18]. After 500 days of continuous operation, the mixed media column is far from its exhaustion. Profile samples indicate that the adsorption front in the column is presently in about 10 cm depth and is moving very slowly. Up to 92 % of the copper is immobilised in the upper 4 cm of the column. Below 15 cm, the copper concentrations in the profile sample are about equal to those in the column effluent (Figure 3). All particulate copper and copper adsorbed on particles with a size larger than 0.45 μm are retained in the upper 4 cm of the media. Therefore, it is suggested that both filtration and adsorption are responsible for the high removal efficiency.

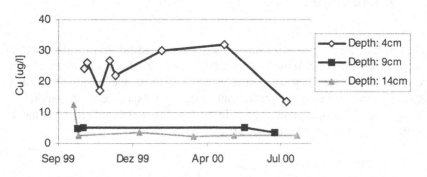

Figure 3. Cu concentration profiles of the FeOOH/CaCO₃ column.

For a copper load of 904 mg, a capacity of 20 mg Cu/g FeOOH can be calculated in the upper 4 cm of the column. However, in order to estimate maximum adsorption capacity, batch experiments were carried out.

In order to gain insight into adsorption kinetics, experiments with a much smaller adsorption bed showing early and distinct breakthrough curves are required. Therefore, two columns with a filter bed only 5 cm high were tested. One was operated as a control column while the other served for kinetic experiments. The goal of the kinetic experiments was to produce breakthrough curves, from which adsorption capacity and kinetic data can be calculated. Again, the inlet copper concentration was the same as for the other columns. After 5 days of steady flow conditions, a steady copper concentration in the effluent could be observed. Then, the hydraulic load was increased in different time steps but with a constant copper load in each time interval. At the end of each time step, a sample was taken in the effluent. As a result, even with a hydraulic load larger than 20 times the constant flow, 80% of copper can be retained in a column with only 5 cm bed length. The residence time in the layer at maximum flow velocity was 20 seconds. After 20 days of constant flow, the same experiment was repeated and again, the same retention properties were found.

5.3. BATCH EXPERIMENTS

5.3.1. Introduction

Because of the rather unstable mechanical properties of the granular FeOOH, shaking of the batch experiment bottles destroyed some of the granules. It was therefore not possible to perform experiments under identical conditions. In order to work with an identical FeOOH suspension, it was necessary to grind the granules. Clearly, this preparation procedure may change the pore size distribution and therefore probably the adsorption behaviour, but this disadvantage has to be accepted in order to carry out reproducible batch experiments. To avoid copper precipitation, pH was adjusted to 6.5. The ionic strength has been set to 0.001 M. As reaction vessels, PE centrifugal tubes with a volume of 50 ml were used. After remaining for the reaction time in an overhead shaker, samples have been taken directly from the tubes and were filtered (0.45 μm) and analysed.

5.3.2.   Kinetic Experiments
The crucial parameters are the ratio of adsorbent to adsorbate and pH. Therefore, the experiments were carried out with two different FeOOH concentrations. Because fast reaction rates were suggested, samples were taken at the beginning in short time intervals. Two series were carried out, each with experiments in parallel. After 5 minutes, 95% of the initial copper in both series was adsorbed onto the FeOOH particles. Equilibrium was reached after 22 hours.

5.3.3.   Adsorption Isotherms
The results of the kinetic experiments show that adsorption kinetics is fast and equilibrium is reached after 22 hours. Therefore, the adsorption process was allowed to reach an equilibrium at least for this time period. Samples with the same amount of iron were prepared while changing the copper concentration in each tube. After reaching equilibrium in the required time, samples were taken and analysed. By calculating the concentration difference, the amount of adsorbed copper can be determined. Saturation can be observed and isotherms can be calculated according to Freundlich or Langmuir. It is important to distinguish between surface precipitation and adsorption. If an equilibrium can be found, it is suggested that adsorption and not precipitation is the crucial removal process. From the isotherms, a FeOOH loading capacity for the inlet concentration of the column of 26 mg Cu/g FeOOH can be calculated. Compared to the actual loading capacity observed in the first 4 cm of the column of 20 mg Cu/g FeOOH, the results are in good agreement. But it has to be stated that there may be a difference between natural roof runoff and pure laboratory systems as well as between FeOOH suspension and the granules.

5.4. FULL SCALE APPLICATION: THE COPPER FAÇADE IN BERN

5.4.1.   Introduction
In Bern, Switzerland, the Swiss Federal Office of Measurements (UFMET) constructed an extension building. The façade of this building, with an area of 2500 m^2, is completely covered with pre-oxidised copper. Because of environmental concerns, a solution should be found to fix the copper in the façade runoff water before infiltrating in the underground around the building. In co-operation with UFMET and the Swiss Federal Office for Buildings and Logistics, it was decided to dispose of the façade water in an infiltration trench constructed around the building and filled with the iron-hydroxide/CaCO$_3$ mixture. In order to test the performance of the adsorption layer, sampling stations were installed at two differently exposed sides

of the building. At each side, special equipment was assembled to catch water from a defined part of the façade and from the effluent of the adsorption layer. The retention efficiency for copper corrosion products can be calculated comparing the copper concentration in the façade water and in the effluent water from the adsorption ditch for each runoff event. All systems are controlled automatically and are able to take samples in very short time intervals. This experimental arrangement permits observation of the supposed dynamics of façade runoff events.

The adsorption layer is situated directly under the end of the copper façade. Its depth and width are 60 cm and 25 cm, respectively. At the bottom, a layer of gravel has been placed. The adsorption layer consisting of a 1:1 mixture of FeOOH and $CaCO_3$ granules is located above the gravel. Its height is 30 cm and it is completely wrapped in a geo-textile as can be seen from Figure 4. In addition, the wrapped adsorption layer is covered with especially structured gravel for aesthetic reasons.

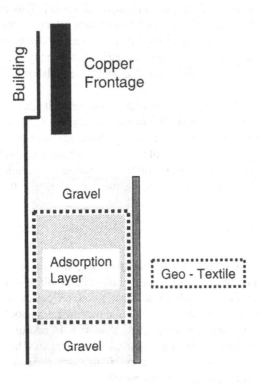

Figure 4. Construction details of the adsorption layer.

5.4.2. Results and Interpretation

Unfortunately, the present results of the copper concentrations in façade runoff as well as the copper concentrations in the adsorber effluent cannot be published yet. According to co-operation agreements with our partners, results will be presented in

178

autumn 2001. Nevertheless, some efficiency data may serve as a first indication of the adsorber performance.

The copper façade was installed in November 1999. The infiltration ditch as well as the sampling stations were put in service in December 1999. Until the end of June 2000, up to 18 and 7 rainfall events including more than 4 samples per event could be sampled from the west and south façades, respectively. The runoff is strongly dependent on meteorological patterns. According to the data, only 4 – 6% of the measured rainfall leads to a façade runoff.

The copper concentration in the effluent of the corresponding infiltration ditch is much lower than the inlet concentration. Comparing the corresponding concentrations from frontage runoff and adsorption ditch effluent, the lowest efficiency was 92% in one event, but otherwise the copper retention efficiency reached 95-99.7% for the all other rainfall events. However, it must be mentioned that until the end of these measurements, the infiltration ditch and therefore the experimental adsorber trench too, were exposed to full solar radiation, because the infiltration ditch was not yet levelled with surrounding soil. This caused big thermal fluctuations which may cause dry out and re-moisture effects of the adsorption layer. Therefore, the structure of the adsorption layer, especially the granular iron, may be affected and consequently, the adsorber performance may be decreased. After September 2000, the terrain will be levelled and better adsorption performance it is expected.

According to the presently available database, it can be stated that the FeOOH/CaCO₃ adsorption layer is able to remove copper highly efficiently from the façade runoff, even if the copper load is very dynamic. Further, it is suggested that both filtration and adsorption take place in the adsorption ditch. To gain more information concerning the copper corrosion process, adsorption efficiency and mechanical stability of the adsorption layer, the experiments will be continued until December 2001.

5.5. DISCUSSION

Although there remain to be some open questions, the tested mixture has shown excellent performance in laboratory and full-scale applications. The adsorption kinetics seem to be fast enough to achieve good efficiency even at high flow velocities. However, more experiments with higher flow rates should be carried out in order to gain insight into process limiting mechanisms. In addition, it is important to understand the kinetic limitations of the adsorption process. As the present data show, film diffusion seems to account for the crucial limitations. Research on this topic is in progress and results are expected in a few months.

The construction layout of the full-scale adsorption ditch in Wabern/Bern can be considered as highly successful. Although its design is simple, the adsorber ditch seems to fulfil functional as well as aesthetic requirements. The question of material stability can be identified as a crucial point remaining to be investigated during long-term operation.

6. Copper and other Heavy Metals in Roof and Street Runoff: Starting Projects

6.1. ROOF RUNOFF

The FeOOH/CaCO$_3$ mixture will be tested with runoff from other roofs. A full scale application of the adsorption layer for treating the runoff from a tin plated copper roof and a titan/zinc plated copper roof is planned. While measuring dynamically influx and effluent metal concentrations, the removal efficiency can be calculated. In addition, to prevent the adsorption layer from clogging, a textile filter mat is inserted before runoff enters the layer. Measurement will allow us to distinguish between retention efficiency of the filter mat and the adsorption layer [19].

6.2. STREET RUNOFF

In another experimental layout, the performance of the FeOOH /CaCO$_3$ mixture will be tested for the removal of heavy metals such as zinc, cadmium, lead and copper in street runoff [20]. Therefore, runoff from a busy road will be collected and conveyed through FeOOH /CaCO$_3$ columns. Again, the road runoff is first filtered through a filter mat in order to avoid rapid clogging of the lower adsorber layer. Removal efficiency of the filter mat and the adsorption layer will be measured and documented.

7. Conclusions

A mixture of granulated FeOOH and CaCO$_3$ has been tested in lab and full-scale applications to remove copper from roof and façade runoff. In order to test the performance, experiments were carried out with roof runoff water in a pilot plant consisting of several columns. A removal efficiency of 99.3% could be achieved with a theoretical copper and hydraulic load of a 20-year return period. Even with this load, the column is far from its exhaustion. To understand the crucial physical and chemical processes involved, experiments with variable flow velocities as well as modelling work are in progress.

The good performance achieved in the column experiments with the FeOOH/CaCO$_3$ mixture encouraged its application in full-scale tests. In these tests, the runoff of a copper façade is percolated through an adsorption ditch. According to present data after 7 month of operation, an overall removal efficiency for copper of 95-99% per rainfall event could be achieved. Although this database is still small, the performance of the adsorption layer can be assessed as excellent. Further measurements will continue until the end of 2001. From a practical point of view, the construction design of the adsorption layer can be considered as an appropriate solution fulfilling functional and aesthetic requirements.

In addition, other materials such as CaCO$_3$ and quartz sand were tested in pilot plants. The results concerning their adsorption capacity for copper are rather poor. These materials cannot be recommended for application in newly constructed infiltration shafts and pits.

180

8. References

1. Swiss Agency for Environment, Forests and Landscape (SAEFL) (1995) Métaux dans les sédiments. Cahier de l'environnement No 240, Berne, CH.
2. Eidgenösische Forschungsanstalt für Agrarökologie und Landbau, Zürich Reckenholz (FAL) (1997) Flächenbezogene Bodenbelastung mit Schwermetallen durch Klärschlamm. Schriftenreihe der FAL 23.
3. Boller, M. and Häfliger, M. (1996) Verbleib von Schwermetallen bei unterschiedlicher Meteorwasserversickerung. Gas, Wasser, Abwasser, 12.
4. Swiss Agency for Environment, Forests and Landscape (SAEFL) (1999) Wohin mit dem Regenwasser?
5. Gysi, U. (1990) Bödenökologie, Georg Thieme Verlag Stuttgart, New York.
6. Boller, M. (1997) Tracking heavy metals reveals sustainability deficits of urban drainage systems, *Wat. Sci. Tech.* **35** (9), 77-87.
7. Mottier, V. and Eugster, J. (1995) Versickerung von Meteorwasser. Swiss Federal Institute for Environmental Science and Techology (EAWAG).
8. Steiner, M. and Boller, M. (1997) Literaturstudie: Untersuchungen zur Eignung poröser Materialien zur Adsorption von Verunreinigungen bei der Versickerung von Meteorwasser. Swiss Federal Institute for Environmental Science and Techology (EAWAG).
9. Lothenbach, B. (1996) Gentle soil remediation: immobilisation of heavy metals by aluminium and montmorillonite compounds. Ph.D. Thesis Swiss Federal Institute of Technology Zuerich (ETHZ).
10. Benjamin, M.M. (1983) Adsorption and surface precipitation of metals on amorphous iron oxyhydroxide, *Environ. Sci. Technol.* **17**, 686-692.
11. Smith, E.H. (1996) Uptake of heavy metals in batch systems by a recycled iron-bearing material. *Water Research* **30** (10), 2424-2434.
12. Benjamin, M.M., Sletten, R.S., Bailey, R.P., and Bennet, T. (1996) Sorption and filtration of metals using iron-oxide-coated sand, *Water Research* **30**, 2609-2620.
13. Edwards, M. and Benjamin, M.M. (1989) Adsorptive filtration using coated sand: a new approach for treatment of metal bearing wastes, *J. Water Pollut. Control Fed.* **61**, 1523.
14. Gabaldon, C., Merzal, P., Ferrer, J., and Seco, A. (1996) Single and competitive adsorption of Cd and Zn onto a granular activated carbon, *Water Research* **30** (12), 3050-3060.
15. Bailey, S.E., Olin, T.J., Bricka, R.M., Adrian, D.D. (1999) A review of potentially low-cost sorbents for heavy metals, *Water Research* **33**, 2469 – 2479.
16. Wagenhaus, G.A. (1991) Crystal chemistry of oxides and oxihydroxides, in D.H. Lindsley (ed.) *Oxide Minerals and Magnetic Significance*, Mineralogical Society of America, Washington, pp. 11–68.
17. Leckie, J.O., Benjamin, M.M., Hayes K.F., Kaufmann G. and Altmann, S. (1980) Adsorption /coprecitpitation of trace elements from water with iron oxyhydroxide, EPR1 RP-910-1, Electric Power Research Institute, Palo Alto, CA.
18. Benjamin, M.M. (1978) Effects of competing metals and complexing ligands on trace metal adsorption at the oxide/solution interface, Ph.D Thesis, Stanford University, Stanford, CA.
19. Ochs, M. and Sigg, L. (1995) Versickerung von Meteorwasser. Swiss Federal Institute for Environmental Science and Techology (EAWAG).
20. Morrison, G.M.P., Revitt, D.M., and Ellis, J.B (1990) Metal speciation in separate stormwater systems. *Wat. Sci. Tech.* **22** (10/11), 53-60.

EVALUATION OF IMPACTS OF CONTROL MEASURES, APPLIED IN THE SOURCE CATCHMENTS, BY MATHEMATICAL MODELS

E. ZEMAN & J. SPATKA
DHI Hydroinform,a.s.
Na Vršich 5
100 00 Praha 10
Czech Republic

1. Objectives

The use of mathematical models for evaluation of flood protection measures and/or flood forecasting has now become a widespread and standard approach. In recent times, there has also been a trend to utilise the results of mathematical models for evaluating the impacts of source control measures both in cities and in rural areas. These measures are being applied in hydrological source basins.

In this case we are dealing with an effort to influence rainfall-runoff processes by measures which can be called source control measures. Proposed source control measures should delay and/or minimise the actual discharge and volume of runoff in upper parts of the basin, in an attempt to reduce flooding in the lower part of the catchment, which is often heavily urbanised. The general principles of these measures are known, but the detailed effects of introducing such measures on runoff parameters are rather hard to measure in a quantitative manner at a particular location by the rule of thumb and/or by empirical formulas.

There is also usually another problem. We need to reduce inflow from the upstream part of the catchment to already highly urbanised areas. But the areas where source control measures could be implemented are located outside the city area and often in areas belonging to other jurisdictions, where the city authorities have no direct political power to enforce source control measures. For these reasons we need a standard method for evaluation and a clear demonstration of how such source control measures will influence the runoff parameters in all their complexity. It is the opinion of the authors that the application of deterministic simulation models can provide the required results, provided that data from a regular or temporary monitoring system are available.

The authors give an overview of the model set-up in the Morava basin, where disastrous floods occurred in July 1997. A source control evaluation formed part of this study. The contribution provides examples and results of application of the suggested methodology.

J. Marsalek et al. (eds.), Advances in Urban Stormwater and Agricultural Runoff Source Controls, 181–194.
© 2001 *Kluwer Academic Publishers. Printed in the Netherlands.*

182

2. Introduction

Water resources assessment involves determining the quantity and quality of water resources, on the basis of which we can make an evaluation of potential actions for development, management and control of water resources.

If we concentrate only on one selected process or just on surface or groundwater flow, the relevant modelling tools are ready for use. We have rainfall – lumped runoff conceptual model and traditional 2D groundwater models available.

In cases where surface and groundwater interaction is rather important, a more comprehensive approach is required. There is clearly a growing need for more modelling applications for water resources assessment.

The main constrains in this respect are counter-productive administrative procedures, sector traditions and the limited user- friendliness of mathematical models.

Abbott and Refsgaard [1] provide a general table of the status of applications of models for various categories of water management problems (Tab. 1).

Table 1. Status of applications of mathematical modelling in several selected fields [1].

Field	Scientific basis	Scientifically well tested	Applied on pilot schemes	Practical application	Major constrains
Water resources assessment - GW	good	good	adequate	standard	administrative
Water resources assessment – Surface water	very good	very good	adequate	standard	administrative
Soil erosion	fair	fair	very limited	null	administrative
Surface water pollution	good	good	adequate	some sub-areas	administrative
Ground water pollution – point source	good	good	partially	standard	techno/ administrative
Groundwater pollution non-point	fair	fair	very limited	very limited	techno/ administrative
Effect of land use changes – flow	good	fair	fair	very limited	Science
Effect of land use change- WQ	fair	fair	fair	null	Science
Aquatic ecology	fair	fair	very limited	very limited	science/techno
On-line forecasting – river levels/flows	very good	very good	adequate	standard	None
On-line forecasting – surface WQ	good	good	adequate	standard	data/admin
On-line forecasting groundwater heads	very good	very good	partially	very limited	data/technol.
On-line forecasting ground water –WQ	fair	fair	null	null	Science
Effects of climate change on flows	good	good	fair	very limited	Science
Effects on climate change – WQ	fair	fair	null	null	Science

Price [2] provides a forecast that by the year 2050 as much as 80% of the world's population will live in urban areas. This has serious implications for society and also for our approach to urban drainage. It is clear that sustainability must be implemented in an appropriate way and on a large scale, and not only in urban areas where the fight for the space will be serious. The sustainability of our approach will be seriously tested during the coming 50 years. The management of water in conurbations must deal with the increase in water supply demands, and storm and waste water collection. Proper management of water will become more complex and more integrated in the future. Urban planners have to understand that they must plan for larger areas and for a greater complexity. Information technologies have brought benefits to the water industry as a whole, but in urban drainage the real benefits are yet to come.

The paper will underline that new trends in urban drainage must go hand in hand with further developments in hydroinformatics, and this is a point-of-view that may be completely new to some engineers, particularly to some well-established industrial experts.

3. Source control measures

Source controls have been defined in many ways, but a very important division line differentiates clearly between urban areas and rural areas. Integration of all processes in urban areas and treating them interactively is the first step of the integrated approach.

We are witnessing another step toward the integration in the current period, because urban planners have to recognise that they must plan and take into consideration a larger area than they are immediately concerned with. The reasons for this change lie not only in heavier demands on water resources that are located far from users, but also because in some cases, it may be cheaper to introduce a source control strategy at a remote location, in order to limit very costly solutions in densely populated cities.

City planners face a problem with respect to deciding exactly when to seek control measures outside the borders of the city. We can recognise four categories of city locations within the "natural drainage system".

The topography of urbanised areas and the type of sewer drainage system used in the city are two very important features that underline how important the categories mentioned in the Tab. 2 are for the introduction of source control measures. We should admit that our categorisation of urban areas may oversimplify a complex matter. Indeed, there are urbanised areas where more then one category are combined (Ceske Budejovice is a typical example of a combination of the 2nd and 4th categories). This brief overview directs our attention toward the location of source control measures. There are no alternatives in the case of the 1st category, to the application of control measures inside the territory of the city. The 2nd and 3rd categories need an individual approach and a remote area for source control measures may be important. The 4th category calls for a flood management evaluation study, in which the impact of several measures should be evaluated along the main river and the impacts of flooding on the drainage system should be considered in alternatives. For this reason it is very important to know whether we are able to translate all source control measures into parameters of deterministic models. Ostrowski [3] introduced several interesting inputs for varying the parameters in deterministic simulation models. If we really are able to produce adequate deterministic models, experts in hydroinformatics will have a major role in the future of source control measures.

Table 2. Classification of cities in relation to their location in a hydrological basin

Category	Location of Urban areas	Danger of flooding from local storm	Danger of flooding from a regional or remote storm
First category	City is located in the upper reaches of the basin (e.g. The City of Svitava)	possible	impossible
Second category	City has a limited basin upstream, and the urban drainage system is of the same order as the rural system (The City of Jihlava on the Jihlava)	most important	exceptional
Third category	City is located in the central area of the basin, and the city drainage system is not comparable with the river drainage system (as Ceska Lípa on the Ploucnice river)	major	major
Fourth category	A large city is located on a large river system, and local flooding has completely different dimensions from flooding by the river system (Prague on the Vltava, Cologne on the Rhine)	important only locally	flooding from the river is a major threat

3.1 SOURCE CONTROL MEASURES IN URBAN STORM DRAINAGE SYSTEMS

Source controls in our present day understanding include non-structural, semi-structural, and structural measures designed to reduce runoff flows and pollutant loads before they enter the drainage or river system. Marsalek [4] describes the control measures listed in Tab 3.

Stormwater management usually involves a combination of source control measures, collection system controls and storage/treatment. Most of the source controls applied in urban drainage can be translated into deterministic models. The changes influence the behaviour of the modelling tools in the form of changed output results of variables (levels, discharges, and velocities). The impact of control structures is evaluated at the suggested location within the current status of the drainage system under consideration.

Table 3. Source Controls for Urban Stormwater Management

Source control	Possibility to simulate/type of model	Description of control measures
non-structural controls	BMP has to be further specified for translation into models/conceptual or 3D physically distributed hydrological models	urban development and resource planning
	YES/ 1D modelling – co-operation - Translation through boundary condition into 1D model	natural drainage
		sewer ordinances and discharge permits
	Hard to simulate	chemical use control
	Hard to simulate	surface sanitation
	Yes - but the data has to be rather complex for calibration and verification purposes	erosion and sedimentation control
structural and semi-structural controls	Yes/1D hydrodynamic model	on-site storage
	Yes/either through boundary condition, or as a structure	infiltration facility
	Yes – but extensive data are required for calibration and verification/conceptual model, Unit Hydrograph, SCS method or physically distributed models	overland flow modifications
	Yes during design/3D HD modelling	solid separation

Both structural and non-structural measures applied in urbanised areas are introduced at a location with known parameters and characteristics. Currently, monitoring networks are denser and more frequently used in urban areas. Data describing an urbanised area are more frequently updated, and there is a widespread availability of thematic maps, DMT and DEM, city development plans and area photographs. Water utilities have most of the data for the drainage system available in their databases or GIS.

This means that placing source control measures into the existing urban drainage set-up of a modelling tool may be a straightforward activity. This approach may basically solve the problems of a city in the first category.

For cities located on low land and ranked in the 2nd or 3rd category, the idea may arise of introducing source control measures outside the city cadastral area, in order to compensate for the increased imperviousness due to new construction activities.

The effect of environmentally-sound measures (such as changes of land use, increase of infiltration, changes of biota) is evaluated by a methodology that was tested in the Morava basin in the flood management master plan during the period from 1997 to 2000. This paper focuses on the maximum exploration of deterministic 1D, 2D and 3D hydrodynamic models of surface or subsurface flow, which allows the user to apply "if, ... then ...." principles during feasibility studies. Since source control measures are quite expensive to implement (e.g., changes of land use), the effect of these measures has to be predictable within a certain range of accuracy.

3.2   SOURCE CONTROL MEASURES IN RURAL AREAS

Marsalek [5] described the interactions in Urban Water Systems by specifying all the relations among the elements of urban drainage in graphical schemes. It is well known that rapid concentration of runoff on impervious surfaces, together with hydraulic improvements of drainage, results in increased incidence and magnitude of floods in urbanised areas. In combined sewer systems, the interaction among the three components is even stronger than in a separate system. When the capacity limits are exceeded in the sewer drainage system, flows are either discharged directly to streams (through combined sewer overflow, CSO, structures) or enter CSO control facilities, which also interact with sewage treatment plants (STPs). Flooding in an urban area may be caused by high surface runoff originating in the city area, outside the city area (from a neighbouring rural catchment), or by flooding through receiving waters. The potential risk of flooding in the city area can be lowered by source control measures inside or outside the city boundaries, in order to compensate for the existing hydraulic improvements. Once a proposal for source control measures in the upper reaches or in rural areas has been made, decision-makers have to know the impact of these measures on the area of interest.

Care for arable land and forests has declined considerably in some countries in recent decades, and it is highly desirable to find a new role for small farmers on their land. The owner of land or forests can be encouraged by grants and/or obliged by legislation to take measures to increase the natural water retention of the landscape, and to decrease rapid runoff through structural measures. Such practices, which are actually source control measures, include:
- changing arable land into forests or meadows
- applying best agriculture and forestry practices in order to increase natural water retention capacity
- creating retention paths on higher slopes
- creating new ponds and polders
- creating natural sediment traps
- making improvements to natural creeks and small rivers, and
- avoiding the use of heavy machinery for harvesting.

This approach gives a new meaning to the activities of landowners in rural areas, provides rural employment, and maintains rural populations.

3.2.1   *Deterministic models of a rural area for source control applications*

Deterministic models are usually based on physical properties of the elements that represent or influence the process under investigation. Three types of deterministic models can be found in the literature:
- DL – Deterministic models based on hydrodynamic laws
- Deterministic conceptual models
- Deterministic black box models

Deterministic models are more suitable for evaluating the impacts in a rural basin that will influence rainfall/runoff processes. Deterministic models have at least basic physical parameters, which can be adjusted in the model either fully, in a descriptive way, or in a schematised, empirical way. Deterministic hydrodynamic models reflect the relation between the cause and effect through processes that determine the relations among variables.

3.2.2 Lumped deterministic conceptual models

Traditional hydrological models of the lumped conceptual type are well suited to deal with the main part of the current water resources assessment, flood forecasting and flood management. Models of this type are highly suitable for the situation where no changes are introduced in the basin, which of course influences the runoff processes. It is noticeable, however, that for these source control measures, where there is a general interest in changing land use, roughness and infiltration rate, and in increasing detention and/or retention, conceptual models are not the best solution from the direct approach point of view.

Lumped conceptual models operate with different but mutually interrelated storage representing physical elements and processes in a basin. Some experts call models of this type " a bookkeeping system", that accounts for moisture contents and/or volumes in reservoirs. Because of the lumped description, where all parameters represent average values over an entire catchment (including all types of non-linearity), the description of processes is not based on governing equations, but on semi-empirical formulas, which nevertheless have a physical basis.

The model parameters cannot usually be assessed from field data, but have to be obtained through calibration and verification. The Standford model is a typical example. A well-known urban drainage stormwater modelling tool that has a lumped module is MOUSE NAM (DHI).

Lumped conceptual models are well suited for simulating rainfall-runoff processes. Models of this type have to be well calibrated and checked against long time series, which have to be measured beforehand. Not all structural measures and/or source control activity within the basin can be simply translated into the conceptual model. The translation needs to be specified by a detailed methodology.

Recently these models have been introduced successfully into daily use, and for this reason they have also been used for source control strategy, especially when data are not available for deterministic physically-based models.

3.2.3 Distributed physically-based models

There is a growing need to use distributed physically based (DPB) models, in which most of the processes are described more accurately. A distributed model operates on the basis of better-fitted and physically-based governing equations of the processes involved. Distributed physically-based models give a detailed and potentially more correct description of the processes in the catchment. The models can be applied to hydrological problems of almost any kind, including the changes related to source control measures in an urbanised area. In such a case, a complex model can be used, involving:

- prediction of the hydrological behaviour in the catchment due to anthropogenic changes (such as land use changes, ground water improvements, higher infiltration through an unsaturated zone, etc.)

- prediction of runoff from an ungauged catchment and from a catchment with only a short time series record,
- water quality and soil erosion modelling.

Models of this type can in principle be used even for an evaluation of changes proposed in the catchment, because the parameters of the model tend to be physically based. We may therefore try to guess the parameters directly from a description of the changes in the catchment.

However optimistic we may be about DPB models, there are constraints and data requirements that have so far prevented DPB models from gaining wider applicability.

Data availability. A prerequisite for making full use of distributed physically-based models is the existence and easy accessibility of a large amount of data, including spatial information and natural parameters such as catchment geology, soil and vegetation, and anthropogenic impacts such as artificial infiltration, agricultural practices, changes of land use, etc. Unfortunately, such extensive data sets are unlikely to exist in the scenarios of cities in the second and the third categories. Furthermore, complex hydrological models also need other types of data, which are held by other authorities (forestry, individual landowners, agricultural data resources, soil physics, etc.). Thus hydrological distributed models may have all the features to solve the problem – "what will happen, if we introduce ....", but need extensive data that either do not exist or are inconsistent with other data sets. Latest developments in remote sensing techniques and GIS give us optimism for the future.

Constraints in the application of distributed models. Refsgaard and Abbott [1] provide an overview of constraints in the application of distributed models in water resources. They listed scientific shortcomings in process description. Only a few engineers and managers are educated and trained with the necessary integrated view of hydrological processes in their physical complexity. The application of distributed models in urban areas or in an adjacent basin, where control measures could be used, often cannot be applied because of incomplete and insufficient data.

One of the most complex DPB models is MIKE SHE (DHI). Kuby [6] presented the practical application of this model for source control assessment elsewhere in these proceedings.

4. Methodology for the Rural Catchment Source Control Assessment

From the reasoning above it is clear that, mainly for urban areas in the second, third and fourth categories, where the user would like to apply deterministic models, there is no "best" solution.

When applying complex DPB hydrological models for the source control assessment outside the urban area, the user will face the lack of data for setting the model up, and also for calibration and validation. The data collection campaign for a full-scale model is also very expensive and time consuming.

When a lumped model is applied for a source control strategy in a rural catchment, we need a special methodology. It may be very misleading to change the already calibrated conceptual parameters by informed guesses.

In the following paragraphs we present a methodology for using a lumped conceptual model based on application of the Unit Hydrograph and SCS methods with CN curves, which can be applied for evaluating source control measures proposed in rural areas for cities in the 2nd –4th categories.

- data collection for a lumped model (rainfall, evapotranspiration data, discharge at selected cross-sections, catchment characteristics, land use),
- data collection for a hydrodynamic 1D model (cross-section data, hydraulic structures, boundary conditions – levels and discharges at boundaries of the domain),
- setting up the MODEL as a combination of a hydrodynamic 1D model and a connected lumped model for the existing situation,
- calibration of the MODEL against the available data,
- verification of the MODEL
- suggestion of source control alternatives
- calibration of the SCS model for areas where the changes will be applied,
- translation of the suggested changes into the SCS model parameters through CN curve numbers – the resulting hydrographs,
- setting up lumped model parameters for the suggested changes in accordance with the resulting hydrographs
- simulation by the MODEL leading to suggested alternatives for design boundary conditions
- evaluation of the source control measures in key parameters.

This methodology represents the best available solution for evaluating source control over large areas when there are no data available for a DPB approach.

DPB models may be developed within the implementation of source control measures, because ongoing monitoring of the impact of the suggested measures is necessary.

5. Case Study of a Conceptual Approach – Morava Flood Management Projects

The effect of environmentally sound measures (such as changes of land use, increase of infiltration, changes of biota) is evaluated by a methodology that was tested in the Morava basin in the flood management master plan during the period 1997-2000.

Source control measures were suggested as one way of dealing with flood mitigation in the Morava basin, because it was widely thought that better land use practice and some reforestation might provide sufficient protection against disastrous floods in the basin. The team had the task of providing the evidence that would either support such a theory or reject it partly or fully.

5.1 CASE STUDY OF APPRECIATION OF LAND USE CHANGES ON AN OUTFLOW HYDROGRAPH – BASIC DATA

The project was executed within the framework of an international project on "Flood Management in the Czech Republic" [7] on the territory of the Morava river basin. The basin was studied from the upper reaches to the Hodonin cross-section. The project was introduced by Zeman et al. [8].

Hydrological data were supplied by the Czech Hydrometeorological Institute (HMI). Time series of measured values from gauging stations on the Morava and Becva rivers and on the main tributaries provided important information. Daily data from 17 gauging stations in the basin were used for the period 1981 – 1986 and for the flood period in 1997. Rainfall data from 130 raingauge stations in the basin were used for the model, and supplemented by temperature data and potential evaporation data. The largest volume of data was prepared for the 1997 flood situation, which is the best documented flood of all times in the Czech Republic.

The total basin, 9700 km^2 in area, was divided into 40 sub-catchments to create rainfall runoff models for each tributary with a basin larger than 100 km^2. The runoff in these basins was generated on the basis of information on rainfall, temperature and evaporation. Calculated values of discharges from rainfall runoff models were used as an input into the hydrodynamic model. The NAM module of MIKE 11 was used as a modelling tool for these rainfall runoff models.

5.2 DESCRIPTION OF THE APPLIED METHODOLOGY

We basically used two types of rainfall-runoff models with different types of description to investigate the response of the drainage area to the flood event. The first of these two models was the NAM conceptual lumped mathematical model, which describes rainfall-runoff processes by a system of linear reservoirs, which characterise single components of the outflow from the drainage area.

The second model that was used for the study utilised the Soil Conservation Service (SCS) method with curve numbers (CN) to describe runoff from various land covers.

The drainage area of the Morava river was divided into 40 sub-catchments. Each of these sub-catchment was divided into 3-5 zones according to the altitude (at intervals of 200 metres), in order to distinguish meteorological and hydrological characteristics and to fit the snow melting pattern. Based on long-term time series of rainfall, temperatures, potential evapotranspiration and discharges in the closing profiles of the catchments (delivered by the Czech HMI), model parameters were calibrated. After calibration of the NAM model, the shapes of the hydrographs were simulated. The observed results were fitted quite well through the calibration and verification procedures, and a generally satisfactory match was obtained.

A scenario of land use changes in the catchment was prepared in co-operation with the Union for the Morava River, after some relevant information regarding possible changes in the alluvial flood plain was provided by the Palackeho University in Olomouc. The actual land use in selected sub-basins was taken from the CORINE database (processed satellite images of land cover) and prepared into Tab. 5.

The envisaged land use changes involve converting 15% of the agricultural land into meadows, (10%) into pastures and (5%) into forests.

The descriptions of hydrological processes in the NAM model are based on a consideration of the physical processes taking place. It is not possible to estimate the NAM parameters directly from the catchment characteristics, but the qualitative effect of land use changes in a catchment area is known and can be quantitatively estimated.

The suggested land use changes would have the impact shown in Tab. 6. The effects of changing these parameters were first tested on one of the catchments, of which 49% of the area is at present under cultivation. The maximum changes were applied to all parameters. The catchment, which has a total area of 537.4 km^2 was divided in two parts as shown in Tab. 7.

Figure 1: Layout of sub-basins in the Morava river basin

192

Table 5. Corine database of land use in the domain.

Name of Catchment	Soil	Resid. areas	Indus. areas	Farms	Cult. land	Meadow & Pasture	Deciduous Forest	Mixed Forest	Coniferous Forest	Other (water)
RASKOV1	B	1%	0%	4%	33%	4%	0%	9%	49%	0%
RASKOV2	B	1%	0%	14%	22%	4%	2%	10%	46%	0%
SUMPERK1	B	3%	0%	14%	8%	6%	0%	15%	53%	0%
SUMPERK2	B	1%	0%	20%	3%	2%	0%	12%	61%	0%
SUMPERK3	B	12%	0%	8%	36%	5%	0%	2%	37%	0%
MORAVICANY1	C	10%	1%	6%	46%	4%	1%	4%	29%	0%
LUPENE1	C	5%	0%	4%	47%	6%	1%	13%	24%	0%
LUPENE2	B	8%	0%	10%	35%	4%	0%	10%	34%	0%
MORAVICANY2	B	8%	0%	10%	39%	6%	0%	19%	18%	0%
MORAVICANY3	C	4%	1%	11%	48%	6%	2%	7%	22%	0%
LOSTICE1	C	3%	0%	5%	47%	2%	1%	14%	27%	0%
UNICOV1	B	3%	0%	11%	36%	6%	0%	23%	20%	0%
OLOMOUC1	C	5%	0%	3%	68%	1%	0%	14%	8%	0%
OLOMOUC2	C	4%	1%	4%	31%	10%	3%	14%	32%	0%
OLOMOUC3	C	8%	1%	4%	57%	0%	11%	7%	11%	0%
JARCOVA1	C	3%	0%	14%	10%	10%	3%	26%	32%	0%
KRASNO1	C	7%	1%	17%	14%	7%	2%	15%	37%	0%
KELC1	C	2%	0%	5%	39%	3%	6%	39%	6%	0%
TEPLICE1	C	7%	1%	7%	57%	2%	7%	8%	11%	0%
DLUHONICE1	C	9%	1%	6%	52%	5%	5%	10%	11%	0%
POLKOVICE1	C	8%	1%	4%	61%	0%	0%	10%	16%	0%
KROMERIZ1	C	5%	0%	5%	54%	1%	6%	7%	20%	0%
KROMERIZ3	C	5%	1%	1%	62%	1%	5%	16%	9%	0%
PRUSY1	C	6%	0%	4%	65%	1%	8%	14%	3%	0%
KROMERIZ5	D	7%	4%	0%	64%	3%	20%	1%	0%	0%
KROMERIZ2	C	5%	0%	1%	79%	0%	0%	2%	12%	0%
KROMERIZ4	C	12%	2%	3%	73%	2%	5%	0%	1%	2%
ZLIN1	C	5%	1%	12%	21%	8%	8%	30%	15%	0%
ZLIN2	C	11%	0%	10%	37%	1%	3%	30%	7%	0%
SPYTIHNEV3	C	13%	3%	4%	50%	1%	1%	22%	5%	0%
SPYTIHNEV2	C	9%	0%	5%	61%	2%	4%	14%	4%	1%
SPYTIHNEV1	C	6%	0%	2%	67%	1%	5%	16%	4%	0%
STRAZNICE1	C	7%	0%	9%	46%	5%	11%	19%	2%	0%
SPYTIHNEV4	C	9%	2%	4%	63%	2%	6%	11%	0%	3%
UHBROD1	C	5%	0%	9%	43%	4%	12%	23%	6%	0%
STRAZNICE2	C	10%	1%	6%	64%	3%	10%	4%	2%	0%
VELICKA1	C	4%	0%	9%	51%	10%	14%	11%	1%	0%
STRAZNICE3	C	6%	1%	4%	74%	2%	8%	2%	1%	2%
STRAZNICE4	C	8%	1%	7%	52%	3%	8%	16%	5%	0%
HODONIN1	B	5%	0%	5%	37%	8%	13%	15%	16%	0%

Table 6. Effect of land use changes.

Effect of change	Parameter	Change to forest	Change to meadow
The root zone capacity would increase due to deeper and/or perennial roots	L_{max}	Increase by 10-50 mm	Increase by 5-20 mm
The increased canopy of the trees will lead to higher interception	U_{max}	Increase by 5-20 mm	Increase by 0-5 mm
Reduced drainage may increase retention in areas with small slopes	CQO $CK_{1,2}$	Reduced by 0.01-0.10 Increase by 10-30%	Reduce by 0.01-0.10 Increase by 10-30%

Table 7. Test catchment parameters.

Name	Catchment	Area Km^2	L_{max} mm	U_{max} mm	CQO	$CK_{1,2}$ Hours
A	Unchanged area	530.4	110	15	0.55	24
B1	15% of agricultural land	43	110	15	0.55	24
B2	B1 changed to forest	43	160	35	0.45	32

The present catchment consists of A+B1, whereas the hydrological conditions after the envisaged land use change are represented by A+B2.

A significant reduction in the runoff peaks was observed. During minor rainfall events, the surface storage remains below its capacity and no overland flow was generated after the land use changes.

In order to verify the order of magnitude of these parameters of the lumped NAM model, we adopted the Soil Conservation Service SCS method for flood runoff estimation. A table with different land use types and the corresponding curve

(CN) was prepared and used for simulation of the same flood period as had been used by the NAM model. The runoff generated by a medium storm of 60 mm and a storm of 100 mm, corresponding to the August 1985 event, was calculated in the table below, using the SCS curve numbers corresponding to agricultural land (79) and meadow & forest (66). The reduction of the runoff to 37% and 55% is similar to the reductions found by NAM.

Table 8. Comparison of effect of land use changes for two events from the August 1985 storm

CN now	CN new	Rain (mm)	Runoff now (mm)	Scenario runoff (mm)	Scenario runoff difference (%)
79	66	60	19	7	37
79	66	100	49	27	55

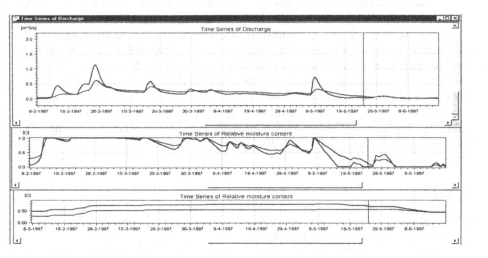

Figure 2 Comparison of two scenarios simulated by the NAM model (full line - current situation, broken line – after changes)

5.3 RESULTS OF THE STUDY

After applying this methodology to the whole Morava river basin, the reduction of the maximum discharge for the historical design flood was calculated at 6-8%, and the main impact was usually seen in the rising part of the hydrographs. These results mean that the maximum extension of a flood source control strategy will produce a Qmax reduction, which is far less than expected by some, who have suggested that approach only, without any other structural measures.

194

6. Conclusion

The authors applied the suggested methodology to evaluate source control measures in a large scale flood management study addressing the effect of source control measures. The authors were able to use the suggested, proven methodology based on a combination of lumped, hydrodynamic models and the SCS method.

Although the evaluation of the source control measures in a rural area was successful, it also showed that such results clearly provide only trends and rough estimates. The suggested methodology can only assist in the evaluation of the measures, if the input data are insufficient. The recommended approach for cities with insufficient data in the 2nd, 3rd or 4th categories is to apply a distributed physically-based model, where all the processes involved are described as accurately as possible. Unfortunately such an approach requires the availability of extensive data.

Since data are rather expensive and can only be provided after they have been collected, the application of a complex DPB model may be impractical.

The recommendation of the team of authors is to start regular monitoring of key parameters for the DPB model in all cases, where source control is suggested outside the city area. Longer than one-year time series with appropriate sampling frequency, including flooding events, will form a basis for DPB models application in source control measure evaluation studies.

The results of the source control study were within the expected ranges. The maximum application of source control measures in the whole Morava basin may lead to a maximum 8% reduction in peak discharges.

7. References

1. Refsgaard, J.C., Abbott, M.B. (1996) *Distributed Hydrological Modelling*, Kluwer Academic Publishers, ISBN 0-792340426, Dordrecht.
2. Price, R.K. (2000) Hydroinformatics and urban drainage - an agenda for the beginning of the 21st century, *Journal of Hydroinformatics* 2, 133.
3. Ostrowski, M.O. (2000) Anthropogenic impacts on the formation of flash floods and measures for their compensation, in J. Marsalek, W.E. Watt, E. Zeman and F. Sieker (eds.), *Flood Issues in Contemporary Flood Management*, NATO Science Series 71, Kluwer Academic Publishers, Dordrecht, pp. 125-133.
4. Marsalek, J. (1990) Current trends in stormwater management and control of combined sewer overflows (CSOs), in *Proc. IAWQ/IAHR Seminar on Urban Storm Drainage*, Prague, Nov. 14, 1990.
5. Marsalek, J., Barnwell, T.O., Geiger, W.F., Grottker, M., Huber, W.C., Saul, A.J., Schilling, W., and Toronto, H.C. (1993) Urban drainage systems: design and operation, *Water Science and Technology* 27, 31-70.
6. Kuby, R. (2000) 3D deterministic modelling in simulation of extreme discharges in urban areas - evaluation of tools for future city development. Preprint, NATO ARW on Source Control Measures for Stormwater Runoff, St. Marienthal, Germany.
7. DHI (1999) *Flood Management in the Czech Republic, MIKE 11 Rainfall-Runoff Modelling of the Morava Catchment Phase Report*, project supported by DEPA Denmark, Horsholm, Denmark.
8. Zeman, E., Biza, P. (1999) Modelling tools - the key element to evaluation of impacts of anthropogenic activities on flood generation, with reference to the 1997 Morava River flooding, in J. Marsalek, W.E. Watt, E. Zeman and F. Sieker (eds.), *Flood Issues in Contemporary Flood Management*, NATO Science Series 71, Kluwer Academic Publishers, Dordrecht, pp. 101-114.

SOURCE IDENTIFICATION CONCEPT WITHIN A FRAMEWORK OF URBAN DRAINAGE MASTER PLANS

KAREL PRYL
DHI Hydroinform, a.s.
Na vršich 5, 100 00 Praha 10, Czech Republic

1. Introduction

The on-going all-society development brings along a fundamental negative effect - a continuous degradation of the environment. Growing urbanization and a denser transportation network change the aboriginal character of the landscape. These substantial changes to the environment force ecological and hydrological processes to change as well. With the change of the ecological and hydrological processes new, usually very specific problems arise. A worsening quality of surface water, groundwater and air quality are amongst the most alarming manifestations of these problems. The level of pollution of these basic environmental attributes then subsequently influences other aspects of life of today's society. The production of wastewater, its treatment, discharge and disposal are all part of an inter-linked system and they altogether form a main factor that determines the quality of the receiving waters.

Master Plans (MP) are essential tools for assessment, maintenance, design or further development of whole systems. Both above mentioned facts, i.e. the need for a good evaluation of the present state of the wastewater discharge from municipal areas regarding present possibilities that the new engineering methods offer, gradually lead to a shift in the philosophical approach to urban drainage. This starts to show even in the Master Plan's execution processes themselves. Former MPs of sewerage and watercourses are now substituted by MPs of urban drainage, which are complex projects based on an integrated approach to drainage of urbanized areas as a whole. The main principle of the contemporary methodology is the complexity, the effort to see each problem as a piece of a more complex puzzle in relation to the other problems, and not to address each particular problem separately, as done many times in the past. Furthermore, we may take advantage of more and more accurate computation capabilities provided by computers.

It is possible to spot the main cause for the conceptual change described above, in the changing approach to the protection of the environment. An optimal solution is obtained only when the complex system of urban drainage, with

J. Marsalek et al. (eds.), Advances in Urban Stormwater and Agricultural Runoff Source Controls, 195–207.
© 2001 *Kluwer Academic Publishers. Printed in the Netherlands.*

196

all the elements shown schematically in Figure 1, is fully evaluated. A complex understanding of the hydrologic cycle in an area is a necessary condition of the integral approach to urban drainage. Integral approach to wastewater drainage and disposal represents the main shift in the conception and evaluation of urban drainage systems.

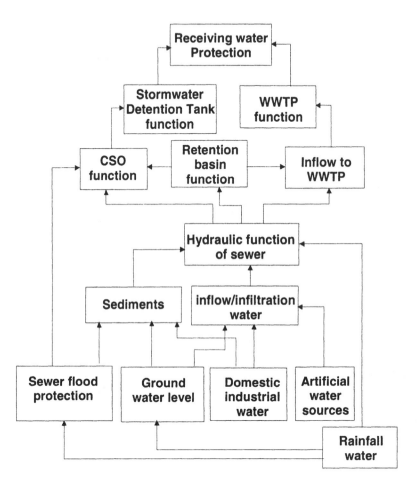

Figure 1. Structural scheme of urban drainage.

2. Master Plan Principles

Fundamental MP principles originate from MP goals and boundary conditions and follow the present urban drainage "philosophy" of high respect to environmental criteria. The sewerage system serves to transport only wastewater with a degree of pollution equivalent to the specified wastewater pollution level so that it could be properly treated at the WWTP. An efficient protection against the inflow/infiltration of "clean water" to the sewerage system is necessary. Surface infiltration of clean water to the ground is recommended with respect to technical and urban possibilities. Otherwise, direct transport to the receiving waters is recommended. The technical measures directed to maximum security and uniformity of the sewerage flows are to be found primarily in a full utilization of the storage capacity or via flow control. In this way local overloading of the system is prevented.

Based on the above-mentioned principles and in accordance with the presently available technologies, four milestone approaches are introduced for the Master Plan Elaboration [1].

2.1. INTEGRAL APPROACH

The integral approach principle originates from the systems view of the drainage of the highly populated, urbanized basins. All the relevant components of the rainfall-runoff process are considered; these include rainfall, surface runoff, transport of media and substances in sewers and receiving waters, biochemical processes, treatment of wastewater in the WWTP, groundwater influence. Based on this, four major elements of urban drainage are pointed out:
1) surface runoff from the urbanized catchment,
2) sewer network and structures,
3) treatment of wastewater, and
4) receiving waters and structures.

2.2. SIMULATION APPROACH

Present technologies, based on the use of mathematical modelling, make possible dramatic quantitative and qualitative changes in the understanding of water related phenomena. Starting with the dynamic, unsteady behaviour of the flow model, taking into account backwater, accumulation, transformation and other effects, through to the advection-diffusion phenomena coupled with biochemical processes, sediment and its transport processes and ending with promising generic algorithms for goal-oriented management of a particular model, these technologies offer a realistic picture not only of the present state of the system, but also of any possible future planning activities or of consequences of extreme and long term events.

198

2.3. DIGITAL APPROACH

The digital way of elaborating the MP is related to the present information technology achievements. Frequently used software - such as CAD programs, databases and GIS along with simulation models - makes it unfeasible to carry out the task in a different way. The advantages of this approach are clear: fast manipulation of large files of data, simple and error free data analysis, easy sorting and selection of information, (relatively) simple transfer of data from one software to other, etc. In this way, maintained MP data are ready for updates or other purposes [2].

2.4. CONTINUOUS APPROACH

The master planning process is a continuous activity, which is to some extent similar to GIS applications. Once the MP is ready, subsequent activities follow. These activities in turn influence the MP solution. Continuity in the update of the MP information keeps the MP "alive" for update or other decision-making processes.

3. MP Goals, Technical Solutions and Working Schedules

Principal political goals could be stated as follows:
   a)   long term strategy for the city drainage development,
   b)   environmental and technical rules application,
   c)   long term investment into urban drainage, and
   d)   digital data source for continuous management of urban drainage.

These political goals could be transformed into the following set of technical tasks:
a)   Quantitative and qualitative function of central WWTP
b)   Drainage strategy in the outskirts of the city and local WWTP
c)   Domestic and industrial pollution
d)   Sewer structures
e)   Inflow/infiltration in the sewer system
f)   Hydraulic properties of the sewer system
g)   Comfort of drainage
h)   Rainfall-runoff reduction management
i)   Flood protection of the sewer system
j)   Impact of sewer system on receiving waters.

To fulfil the above-mentioned goals, a sophisticated and a well-prepared organization scheme has to be established as shown in Figure 2.

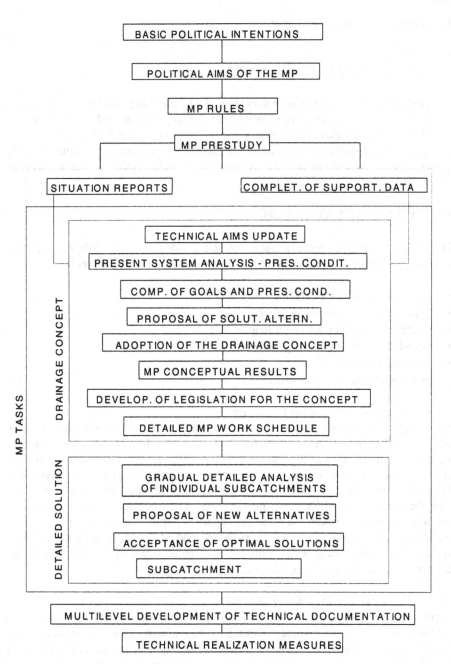

Figure 2. Master plan work schedule.

4. Input Data – Situation Reports

4.1. THE AIM OF SITUATION REPORTS

The goal of a situation report is to document and evaluate the present state (situation) of the most important elements of the whole drainage system and to determine the boundary conditions that are necessary for the creation of the MP. The aim of the situation report thus is not to analyze the behaviour of particular elements of the drainage system or even to describe the processes that take place within the drainage system[3].

4.2. TYPES OF SITUATION REPORTS

The collection of the primary data may be divided into the following groups, according to the particular element of the drainage system the data refers to :

- situation report of the catchment
- situation report on sewer network
- situation report on wastewater treatment
- situation report on receiving waters
- situation report on infiltration
- situation report on ballast water.

4.2.1. Situation Report of the Catchment

The objective of a situation report of the catchment is to provide information about the present state of wastewater discharge from the catchment through the sewerage (or the surface watercourses) and information about possible negative effects of individual objects in the catchment on the operation and functionality of the drainage system. A situation report of the catchment together with other situation reports and materials offer the processor the information that is essential for the analysis and evaluation of the pollutant load to which the drainage system is exposed under various boundary conditions, for the comparison of the pollutant load of particular parts of the sewer system, indication of large water spills etc. A situation report of the catchment is the cornerstone for surface runoff evaluation.

4.2.2. Situation Report on Sewer Network

A precise description of the technical state of the sewer network and its objects and a definition of appropriate boundary conditions that influence the functionality of the sewerage and its objects are the general goals of this type of report. The data from a digital land register, a sample of which is shown in Figure 3, should be supplemented by the information gained through operational experience, and in the preparation of the situation report on the sewer system.

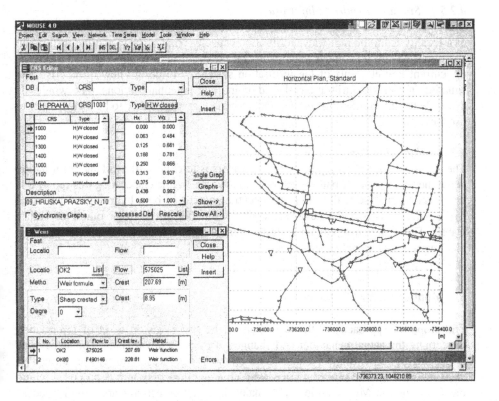

Figure 3. Data on the sewer network.

4.2.3. Situation Report on Wastewater Treatment

The aim of a report on wastewater treatment is again to collect the fundamental information about the structure, technology and aspects of operation of all the WWTPs in the area, including the information about the particular impacts on the receiving watercourses.

4.2.4. Situation Report on Receiving Waters

The aim of a situation report of a watercourse is a perfect description of the watercourse that will be used within the frame of the drainage system. This description means mostly the technical and operation data of the watercourse and its objects. Such data make up the ground for any further reasoning of water management nature considering drainage once the simulation model is set up.

Another very important goal of this situation report is an ecological evaluation of the appropriate watercourse, its pollution level and the effects of the sewerage or other factors on the bio-ecological load. The complex approach to MP elaboration requires many fields to cooperate when proceeding and, unlike in the past, the participation of an expert biologist. The meaning of such expert's participation in the drainage system preparation is not in elaboration of a complex biological study, but in a meaningful help with identification and localization of the pollution sources and in finding the best way how to reduce potential pollution of the watercourse.

4.2.5. *Situation Report on Infiltration*

The goal of a situation report on infiltration is to document and analyse the actual state of a locality from the rain water infiltration capacity point of view and the infiltration conditions (hydro-geological, catchment area) point of view in general and to identify the current key elements of the infiltration system or to design new ones. Also, the situation report on infiltration should point out potential regions that would be available for infiltration together with the criteria that would have to be met in order to use these regions.

4.2.6. *Situation Report on Ballast Waters*

The goal of the situation report on ballast waters is to collect all available information about the total runoff of ballast waters from the area. While elaborating this report it is necessary to analyse real as well as potential sources of both hydrological and non-hydrological waters and to evaluate their contribution to the total runoff of ballast waters. The situation report on ballast waters works with the data from monitoring, pictures and operational experiences.

5. Monitoring for Master Plans

The proposed approach for the execution of the MP relies on a set of information concerning the actual behaviour and changes of the system variables. The actual way of "getting to know" this behaviour based on the in-situ measurements was defined by us as monitoring. The procedure of setting up a monitoring system is proposed within the frame of the MP activities. The main aims of this activity are summarized as follows [4]:

    a)   calibration and verification of the simulation model,
    b)   examination of the actual state of the system,
    c)   proof of the correctness of the executed measures, and
    d)   statistical evaluation of trends.

    Preliminary preparation and establishment of the monitoring system itself is recognised, together with a detailed and frequent inspection, as a key element of any monitoring campaign. To design an optimal monitoring scheme and to suggest a logical and functional system of operation of monitoring sites (including data collection, transfer and evaluation) for large scale projects (such as Master Plans) is a very important conceptual step in project elaboration. Based on the drainage system analyses and in a collaboration with a competent Waste Water Board, having in mind overall project goals, it is necessary to implement the following activities within the monitoring campaign:

- the definition of key parameters, which are going to be measured, including their time and area distribution,
- the definition of the methodology for the measurement of all parameters,
- the location of proper measurement sites, including applied technology, and
- the proposal of a primary system of data collection, transfer, control and storage.

All of these activities could be seen as a preparation stage of the project. It is recommended to pay relevant attention to them, as they could affect the quality of the final outputs of the whole project. On the other hand, they represent rather large amounts of work and finance within the project.

As stated in previous paragraphs, the goals of the monitoring campaign are based on the goals of the whole project and they are closely connected. The major specific goals for a particular period of the project are as follows [5]:

- to get the information about rainfall spatial distribution,
- to select proper rainfall events (or periods, depending on a selected modelling technology) for a calibration and verification of mathematical models,
- to obtain a long term rainfall and outflow time series,
- to set a background for historical rainfall time series, including the space and area distribution,
- to describe the hydrodynamic behaviour of main sewers, based on measured discharges,
- to describe the operation of main CSOs, based on measured water levels,
- to obtain a general record of dry and wet weather flow water quality in sewers,
- to obtain a general record of flow and water quality in receiving waters, and
- to suggest a conceptual long term monitoring strategy for the whole urbanised area based on the obtained data.

Measurements and monitoring represent an essential part of urban drainage MPs. As the measured and evaluated data belong to the most important input parameters for the modelling activities, it is necessary to ensure close cooperation between measurement and modelling teams, in order to select a proper measurement methodology and monitoring sites as well. This cooperation significantly affects an overall success of the whole project.

Regarding the equipment used for such project - it is important to select measurement devices with a high degree of equipment reliability and also to put emphasis on the interfacing of the used equipment (according to the operation and maintenance). To be able to handle such amount of data confidentially, it is necessary to have relevant and capable software tools for data processing. And last, but certainly not least, it is necessary to mention that the efficiency and success of the so-called monitoring activities highly depends on the planning and organization of each measurement campaign.

6. Advanced Simulation Tools and Results Presentation

Simulation models represent a clear evidence of the progressive utilisation of information technologies in the world of classical technical branches. The modern approach to the process of finding technical solutions and elaborating large scale studies in the field of water management is based on new technologies that are built upon still growing possibilities of contemporary computers. The essential elements of this approach - the deterministic simulation models - are such software tools. They take into account cumulative effect of geographical, hydrological, hydraulic, chemical, and other factors in relation to the particular problem. With the aid of these models it is possible to get an accurate image of the behaviour of a particular element of the system (e.g. the sewerage) and then to analyze the effects of suggested alternative solutions in the sense: "Once we realise these changes in the network... then we have to be ready for these consequences." This means that there is a model of a segment of the real world in a digital form within the computer, where we can simulate and evaluate different alternatives without economical or ecological hazards.

On the other hand, it seems reasonable to point out that even though models are a big help in the process of MP elaboration, they are only "handy" tools that do not solve problems themselves and there still is a fair amount of work that has to be done besides using the models.

Figure 4 illustrates potential of relevant outputs from simulation models. All of the presented data were processed within the elaboration of the Urban Drainage Master Plan for the City of Prague.

205

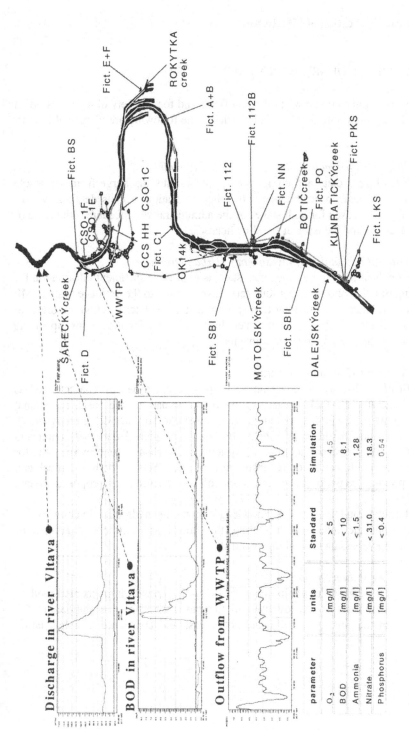

Figure 4. Impact on receiving waters (Q, W_Q parameters) – the Vltava river in Prague.

7. Practical Application of MP Results

7.1. MAIN FIELDS OF MP RESULTS APPLICATION

MP results are applicable in a wide range of fields and for a variety of activities. Most of them are very important for the whole society. The following are of special interest to us.

7.1.1. City Development with Respect to Spatial Planning
The MP lays down a clear and complex concept of water discharge from the whole area of the catchment, in accordance with other legislation. For its detailed way of elaboration, a MP is used for both stages of the administrative procedure – the issue of territorial decision and the issue of building licenses.

7.1.2. Systematic Planning of Investment in Urban Drainage
One of the goals of MP, at the conceptual state, is to provide sufficient information for the planning of either short-term or long-term investment into the drainage system. MP is especially feasible for the long-term investment plan determination in order to achieve the selected quality level of water discharged and for the setting up of the background for the ongoing local investments.

7.1.3. Systematic Operation Activities
The benefit of the MP to the sewer network operator could be best characterised as giving the operator the tool for verification of the impact of the basic operational measures on the sewerage (e.g. relaying, main sewer checkup, accidents recovery...), and for close study of the system behaviour based on gained data evaluation and interpretation. A permanent monitoring system is established to provide data for possible future modification or reconstruction of the CSOs in both technical and economic-political sense or as an indication of either current or proposed system overload [6].

Calibrated models give the knowledge about the hydraulic function of the sewerage, about the manipulation during the floods and extreme situations, or eventually supply the data for real time control.

7.1.4. The Legislation Support
In some cases the current legislation may not be up to date with implemented modern technologies that may be already used in the practice. From this standpoint, the outcomes from the final MP could and actually should contribute to a legislation update.

7.2. POST PLANNING STUDY ISSUES

Once the MP is completely done, it still is not all over. There are further steps that have to be taken with respect to the city development. These are the next phases that follow the MP elaboration [7]:

- engineering design,
- post project appraisal,
- maintenance of models and databases, and
- cost benefit assessment.

8. Observations

Not much to conclude as yet, for the whole article is already a sort of summary. But still, it somehow seems to be appropriate to emphasize at this point that current Master Plans, based on the application of modern hydro-informational tools, are "living organisms", which could answer a lot of practical questions related to urban drainage systems. But on the other hand, in order to keep these Master Plans "alive", it is necessary to allow their continuous update. To turn all these ideas into practice, we have to directly transfer the reality and some future visions into the digital world.

9. References

1. Pliska, Z., Metelka, T., Mucha, A., and Pryl, K. (1998) Pilot project for urban drainage master plan of Prague, 1st.Int.Conf. Master Plans for Water Utilities, Prague, Czech Republic, pp. 65-70.
2. Pryl, K., Vasek, P., and Vanecek, S. (1998) Data processing used for urban drainage systems, UDM '98, London, England.
3. Mucha, A. and Pryl, K. (1998) Urban drainage master plans, *SOVAK*, **7** (9), 2-5.
4. Gusstafson, L.G., Kuby, R., Metelka, T., Mucha, A., and Pryl, K. (2000) The urban drainage master plan for the City of Prague, in Proceedings of the Integrated Modelling User Group (IMUG), April 12-14, 2000, Prague, Session 1, 1-5.
5. Pryl, K., Krejcik, J., and Dolejs, M. (2000) Permanent and short-term monitoring for master plans, in Proceedings of the Integrated Modelling User Group (IMUG), April 12-14, 2000, Prague, Session 4, 1-5.
6. Richardson, J., Pryl, K. (2000) Computer simulation helps Prague modernize and expand sewer system, *WATER Engineering and Management*, **147** (6), 10-13.
7. Foundation for Water Research (1998) *Urban Pollution Management Manual, A Planning Guide for the Management of Urban Wastewater Discharges during Wet Weather*, 2nd edition, Marlow, England.

THE ROLE OF A SYSTEM "HYDROLOGICAL MEMORY" IN SOURCE CONTROL STRATEGY EVALUATION IN URBAN AREAS

T. METELKA & T. LAICHTER
Urban drainage modelling
Hydroprojekt a.s., Taborska 31, 140 16, Prague 4, Czech Republic

1. Introduction

A practical use of surface runoff modelling in urban and natural areas involves several process variables, which highly influence final model results. The selection and good interpretation of rainfall events as well as the evaluation of impervious and pervious areas, and a lag time assessment, belong to the most crucial tasks in deterministic hydrological modelling mainly, if a long-term system behaviour is to be analysed. Unfortunately, the evaluation of these parameters is in most cases not an easy task, because of their spatial and temporal changes [1]. A system "hydrological memory" plays an important role in this process and should be considered in particular catchment modelling, if accurate long-term hydrological behaviour is to be simulated.

A deterministic, conceptual hydrological model MOUSE NAM is applied in the Urban Drainage Master Plan of the City of Prague. This program is capable of long-term simulation of rainfall/runoff processes for impervious areas as well as the simulation of rainfall-generated infiltration from pervious areas to the sewers. This way, a quantitative distinction can be made between a fast runoff component FRC (inflow) and a slow runoff component SRC (infiltration) and the total runoff water balance can be evaluated in catchments [2]. The model also simulates antecedent hydrological conditions (e.g. a land phase of the hydrological cycle) during long term simulations. The results of the simulation can yield an important information concerning infiltration into the sewer system, total runoff from the catchment in the form of inflow and infiltration volumes, and consequently the information concerning the actual treatment capacity of the Wastewater Treatment Plant (WWTP). While a fast runoff inflow can cause local capacity overloads, combined sewer overflow (CSOs) spills and pollution in the receiving water, and a slow runoff infiltration into the sewer system brings about a hydraulic overload at WWTP, a dilution of wastewater and, consequently, the deterioration of treatment processes. These facts highlight the importance of understanding the rainfall runoff processes and their quantification as a basis for proposal of remedial measures.

The experience from the project of the Urban Drainage Master Plan of the City of Prague proves that different hydrological conditions can affect peak flows as well as volumes of the actual runoff hydrographs. The differences in rainfall runoff modelling results can be then observed under different hydrological conditions and hence they should be considered in the final remediation proposal.

J. Marsalek et al. (eds.), Advances in Urban Stormwater and Agricultural Runoff Source Controls, 209–220.
© 2001 *Kluwer Academic Publishers. Printed in the Netherlands.*

2. Study case - Urban Drainage Master Plan of the City of Prague

At present, the Urban Drainage Master Plan for the City of Prague is underway at the Hydroprojekt company in consortium with DHI Hydroinform [3]. The need for the new master plan was generated by the rapid city growth, a need for implementation of EU legislation as well as by the age of the former master plan (more than 20 years). Furthermore, the development of new approaches and technologies in the area of Urban Drainage brought new possibilities for modern master planning in Prague. The project started in November 1998 and the completion is expected by June 2001.

2.1. PROJECT AREA

Prague is the largest urban agglomeration in the Czech Republic. The city population is at present more than 1.3 million inhabitants living in an urbanised area of approximately 500 square kilometres. The city is located on both sides of the Vltava River close to the confluence with the Elbe River. The sewer system is of the combined type and covers about 60% of the total city area. It consists of 2,360 km of sewers, 54,000 manholes, 140 CSOs and 19 pumping stations (see Fig.1.). The sewer system is connected to the central Wastewater Treatment Plant that was designed for a flow rate of Q = 6 m³/s. The morphology of the area varies from steep slopes (near the river) to almost flat areas on the outskirts.

Figure1. Combined sewer system in Prague

2.2. PROJECT APPROACH AND METHODOLOGY

The Master Plan is based on an integrated approach that enables the evaluation of surface runoff as well as the assessment of transport and treatment processes in sewer and river systems. Fully digital execution and the use of mathematical modelling are fundamental for this project. The project is executed in 5 modelling phases starting from the very simplified large-scale phase comprising the whole catchment of the city and gradually coming to a more detailed evaluation in defined subcatchments. These phases are defined according to the project goals and in relation with the task formulation.

Based on the task formulation, the schematisation of the system and processes was carried out for each project phase to fulfil pre-defined goals. In the first modelling phase, the total combined drainage area of 12,000 ha was schematised into 19 large subcatchments for hydrological simulation with the MOUSE NAM model. The sewer system was reduced to main collectors around the treatment plant and simulated with the MOUSE HD, a one-dimensional hydrodynamics and water quality model. The Vltava River water quality was of a main concern in this phase and it was simulated with MIKE11, HD, WQ one-dimensional hydrodynamic and water quality model. The overall system schematisation is shown in Fig. 2.

The main concern of this phase was directed to the evaluation of the overall system functionality, long-term water balance, capacity of and storage possibilities in main trunk sewers, and the actual treatment capacity of the central wastewater treatment plant. The impact of the sewer system and WWTP on water quality in the Vltava river was also evaluated. This contribution refers to the results of the above described project phase.

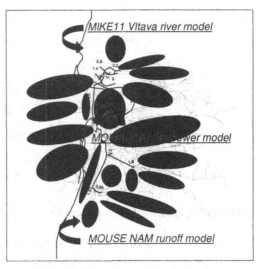

Figure 2. System schematisation in 1st. phase

In the subsequent two phases, each of subcatchments was again subdivided into approximately 50 – 100 subcatchments for simulation. The sewer system was defined in more detail for all main catchments (see Fig.3.), and the selected main tributaries of the Vltava River were also included. The same methodology and technology was applied in these phases for the evaluation of the system behaviour. The main focus of these phases was on the CSOs functionality, sewer

Figure 3. Complete model set-up for selected main catchments

transport and capacity function, and impact on water quality in selected tributaries. These phases are being presently finalised.

The boundary conditions schematisation was defined in relation with the project goals and tasks in a form of a long-term historical precipitation, evaporation and temperature records. Based on this approach, a long-term simulation (from 1 to 10 years – see Fig.4.) of the system was carried out, along with a single extreme event simulations for selected return period rainfalls.

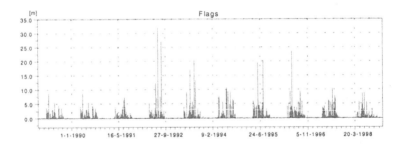

Figure 4. Example of a 10-year historical rainfall series

3. Modelling surface runoff processes

The modelling of surface runoff is composed of modelling several natural processes which have to be analysed. Runoff generation and water collection components of surface runoff process are recognised [1].

- **Runoff generation processes**

Figure 5. Historical rain-gauge locations

Runoff generation process is a highly random phenomenon, which is difficult to model. The use of stochastic methods seems to be appropriate for defining artificial long-term rainfall series. The Prague project team could, however, profit from the availability of measured rainfall data in the project catchment. 10-year continuous precipitation records (1990-1999) are available at 7 rain-gauge stations. They are in a digital form, verified and free from random and statistical errors (see Fig.5).

In addition, a comprehensive monitoring campaign of rainfall is being carried out during the whole project execution (starting from June 1999). 19 heated digital rain-gauges are operated (see Fig.6) with the sampling frequency of 1 minute. The data are regularly collected, evaluated and processed. The gauge station network covers a total area of approximately 400 km^2.

The described precipitation data sets help the project team evaluate the spatial and temporal behaviour of the rainfalls over the Prague area

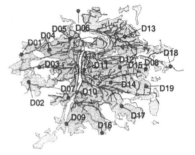

Figure 6. Project rainfall monitoring set-up

and to apply some deterministic approaches for extending point measurements of a spatially variable phenomenon to particular catchment sizes. There are several alternative techniques available for coping with spatial and temporal rainfall distribution over a particular catchment area (Thiessen polygons, Inverse Distance-Weights, Kriging, splines, etc.). The project team selected the Inverse Distance-Weights method to be used for rainfall distribution analysis in the Urban Drainage Master Plan of the City of Prague.

- **Runoff collection process**

The mathematical modelling of the runoff collection process is dominantly a deterministic phenomenon. Thus, a selected deterministic hydrological model can be used to model the surface runoff and to generate runoff hydrograph. A deterministic, conceptual hydrological model MOUSE NAM is applied for surface runoff modelling in the Urban Drainage Master Plan for the City of Prague.

This model is defined by a set of linked mathematical statements describing, in a quantitative manner, the behaviour of the land phase of the hydrological cycle. The model is capable of computing not only runoff from impervious areas (rainfall induced inflow), but also the so-called "rainfall induced infiltration", which in fact does not depend only on the actual precipitation, but is affected by the previous hydrological situation. For a certain rainfall event, the runoff hydrograph will vary, based on the previous hydrological state of the system. According to the response time, the rainfall induced inflow from impervious areas is understood as "a fast runoff

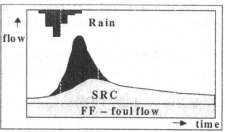

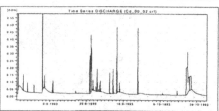

Figure 7. Components of runoff hydrograph

component" (FRC) and the rainfall induced infiltration generated from pervious areas as "a slow runoff component" (SRC). While the FRC response time is counted in the order of minutes (or tens of minutes), the SRC response time takes from hours up to weeks. The calculation of the FRC routing is implemented in a form of either the time/area method or the non-linear reservoir method, while the effective rainfall (net precipitation) is generated as a portion of natural rainfall based on the actual conditions of snow and surface depression storages. This way, a wetting process, evaporation and filling of depressions are accounted for. The model of SRC is implemented in a form of overland discharge, groundwater level, relative soil moisture and baseflow calculations. These calculations take place in four types of inter-related storages: snow storage, surface storage, root zone storage and groundwater storage.

214

3.1. CATCHMENT SCHEMATISATION

The total Prague catchment size is 500 km². Out of this area, approximately 12 000 hectares are drained by the central, combined sewer system. The schematisation of the catchment of this size is not an easy task. The hydrological model MOUSE NAM treats every catchment as one separate unit (like the majority of current surface runoff models). The catchment parameters, therefore, have to represent an average for the whole catchment.

During the first modelling phase of the project, the total area was divided into 19 large size subcatchments (see Fig.8). Each subcatchment was later schematized in the form required by the hydrological model NAM. The average model parameters were

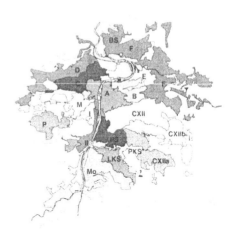

Figure 8. The main subcatchments of Prague catchment

then validated during the calibration procedure. Precipitation data for the simulation had to be modified according to the size of a particular catchment.

3.2. MODEL VALIDATION

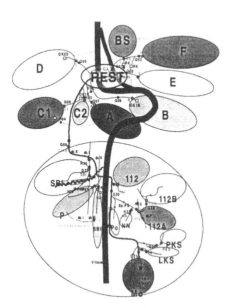

Figure 9. The main subcatchments of Prague catchment

The calibrations and verification of the hydrological model MOUSE NAM were performed together with hydrodynamic calibration at selected sewer locations, in order to validate the model parameters. Together, 19 ADS flow meters are used in the project to calibrate each subcatchment model at the downstream boundary (see Fig.9).

The calibration period was defined according to the monitoring campaign of 10 months (May 1999 – February 2000). This way, MOUSE NAM long-term response parameters could be validated (see Fig.10).

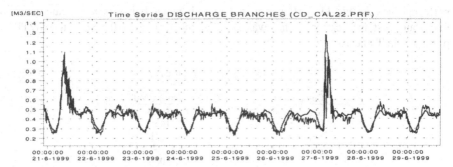

Figure 10. The calibration of the main Prague subcatchments

4. Long-Term System Behaviour

The evaluation of a long-term behaviour of the Prague urban drainage system was considered as one of the most important activities during modelling. Long-term average inflows and volumes of wastewater components (SRC, FRC and foul flow) represent important information with regard to the central treatment capacity in Prague. Based on these data, a principal decision had to be made concerning the future treatment state (either the intensification of the current WWTP or construction of a new one on the outskirts of the city).

The historical precipitation data from the year 1994 were selected as a boundary condition for long-term simulation after the analysis of annual rainfall sums from a 9-year historical set (see above). The selected subset was agreed to represent "an average" precipitation year (annual sum: 445 mm). Other model data (population, water consumption, daily flow variation) referred to the current state of the system in year 1999.

4.1. ANNUAL WWTP INFLOW ESTIMATES

Based on the precipitation year 1994, the long-term simulation of a current system state was performed. The simulation results were gained in a form of one year time series of flows, water levels and velocities at each model location (nodes, pipes, CSOs, pumps, etc.). Using this information, the so-called "flow schemes" were developed for the main trunk sewers to help in the understanding of a long-term

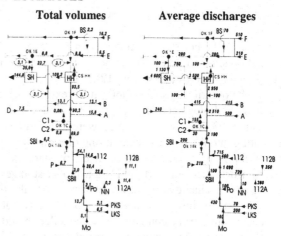

Figure 11. Long-term flow schemes

216

performance (see Fig.11). After that, the overall water balances were calculated at important system locations. The evaluation at the inflow to the central WWTP is presented in Table 1.

Table 1: WWTP annual water balance

Component	Annual volume (Millions m^3)
V $_{total}$ produced	149.8
V $_{total}$ treated	145.2
V $_{SRC}$ treated	42.1
V $_{FRC}$ treated	2.4
V $_{FF}$ treated	100.7
V diverted (CSOs)	4.6

4.2. ROLE OF A SYSTEM HYDROLOGICAL MEMORY

The long-term simulation results gave important answers concerning the wastewater load as well as the current hydraulic capacity of WWTP. However, during the hydrological modelling it was recognized that the antecedent hydrological situation plays an important role not only during the long-term simulation. The use of a particular simulation hot start influences the final simulation results as well. If the historical year 1993 was used for the simulation of a hot start (wet year with a rainfall sum = 530 mm), a substantial increase of infiltration into sewers would be noticed

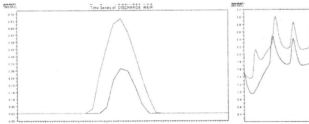

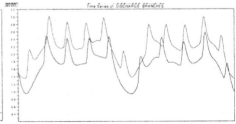

Figure 12. CSO overflow change caused by wet or dry hot start

Figure 13. Dry weather flow change caused by wet or dry hot start

compared to the hot start generated from the "average" year 1994. The influence of a hot start on simulation results can be then observed during both dry weather flows (see Fig.13) and storm events (see Fig.12).

The described phenomenon can be explained, in agreement with the reality, by a system "hydrological memory" included in the NAM model. During the wet end of the year 1993, all model storages were filled up with rainwater, which was then infiltrating into the sewer system during the coming year spring (1994). Once the same year 1994 is used for simulation and hot start, the hydrological situation differs. During a dry year hot start simulation model storages are empty and, consequently, less water is available after the winter season for subsequent simulation.

From the above facts it seems to be obvious that natural hydrological behaviour of a catchment is very complex and a modeller should not set his mind on one set of results after a single simulation. Spatial and temporal variabilities of rainfall-runoff processes bring different response under different conditions. In the Prague case, the hydrological modelling was performed for several hydrological conditions. The simulation analyses then necessitated the presentation of results in terms of ranges of values, rather of discrete values. For instance, the infiltration volume treated annually at the central WWTP is expected to vary between 20 – 50 mil. m^3/year (20-40% of the overall treated volume).

5. Source Identification and Control Measures

The development and implementation of source control measures for rainfall-induced infiltration is not an easy task. On the other hand, a possible reduction of sewer rainfall induced infiltration could bring important improvements in the urban drainage performance. In the Prague case, the reduction of up to 20% of annual treated volumes is expected to be reached at the central WWTP. This would have important consequences for the treatment efficiency as well as for the overall treatment cost savings.

It is obvious from the above mentioned facts that the rainfall induced infiltration is considered to be an important drainage phenomenon, which should be analysed in the present master planning activities. In the Prague case, the analysis of rainfall induced infiltration is being carried out, infiltration sources are identified and control measures proposed as a part of the final remedial scheme for the Prague Urban Drainage Master Plan. Interim project results are presented below.

5.1. IDENTIFIED SOURCES

The annual infiltration volumes generated from all main Prague catchments were evaluated for the simulated year 1994 and hot starts corresponding to the years 1993 and 1994. The results are presented in the table below in the form of annual volumes, where Vtot represents the total volume of wastewater produced in the particular catchment, Vsrc the infiltrated volume, Vfrc the rainfall inflow volume and Vff the foul flow volume (See Tab.2).

It is clear from the above results that the hydrological memory of hot-start years 1993 and 1994 leads to large differences in produced and treated volumes of wastewater. This difference is not only seen in each subcatchment, but mainly for integrated values at the WWTP.

This way, the infiltration source volumes were evaluated in all the defined subcatchments. It was recognised that an important (almost constant) infiltration source is located in the vicinity of the WWTP (catchment REST in above table). This area produces annually 12 mil. m^3 of water which is drained into the sewer system. After the analysis of the size and surface of the area, sewer system topology and the location of the WWTP (on the Troja island on the river Vltava) it was concluded that most probably at least one of the four inverted siphons transporting the wastewater from the main trunk sewer to the WWTP is infiltrating water from the river. This finding is going to be verified by additional quantitative and qualitative measurements at this site.

Table 2. Annual water balances of distinct wastewater sources for defined simulation cases

Catchment name	1994 - 1994				1993 – 1994			
	V tot [mil. M3]	Vsrc [mil. M3]	Vfrc [mil. M3]	Vff [mil. M3]	V tot [mil. M3]	Vsrc [mil. M3]	Vfrc [mil. M3]	Vff [mil. M3]
WWTP	125.50	22.78	6.52	99.15	149.40	42.15	6.51	100.74
A	14.3	0.28	0.84	13.18	15.83	1.7	0.84	13.29
B	11.2	0.6	0.75	9.85	13.20	2.4	0.75	10.05
C1	4	0.25	0.32	3.43	4.90	1.1	0.3	3.50
C2	0.8	0.01	0.15	0.64	0.80	0.03	0.15	0.62
D	5.9	0.37	0.47	5.06	7.54	1.91	0.47	5.16
E	4.1	0.37	0.27	3.46	6.50	2.8	0.27	3.43
F	12.3	0.6	0.89	10.81	16.24	4.4	0.89	10.95
BS	2.1	0.04	0.19	1.87	2.34	0.2	0.19	1.95
REST	12.6	12.6	-	-	12.60	12.6	-	-
SBI	4.9	0.97	0.25	3.68	6.26	1.46	0.25	4.55
P	5	0.18	0.14	4.68	6.70	0.78	0.14	5.78
SBII	2.4	0.25	0.21	1.94	3.00	1.5	0.21	1.29
112	13.5	0.22	0.42	12.86	14.68	1.34	0.42	12.92
112B	9.7	3.31	0.56	5.83	11.22	4.62	0.56	6.04
112A	10.4	0.71	0.31	9.38	11.40	1.6	0.31	9.49
Po	2.1	0.32	0.32	1.46	2.18	1.36	0.33	0.49
NN	0.2	1.03	0.03	0.86	0.30	0.15	0.03	0.12
PKS	1.7	0.1	0.1	1.5	2.11	0.46	0.1	1.55
LKS	5.8	0.3	0.2	5.3	6.50	0.94	0.2	5.36
Mo	4.5	0.27	0.1	4.13	5.10	0.8	0.1	4.20

Additionally, it has been concluded that the rainfall-induced infiltration is also affected by the emergency weir at the WWTP pump station. The spill frequency at this weir dramatically increases under wet-weather conditions (Figs. 14 and 15).

From the displayed time series it is clear that for the 1993-1994 simulation case, even dry-weather flow peak spills over the weir to the Vltava River in the beginning of the year. The total overflow volumes then increases from 1.8 mil. m^3 for the 1994-1994 simulation case to some 3.5 mil. m^3 for the 1993-1994 simulation case.

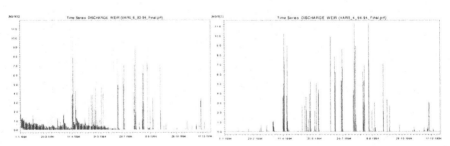

Figure 14. WWTP emergency weir operation for 1 year simulation case 1993-1994

Figure 15. WWTP emergency weir operation for 1 year simulation case 1994-1994

This situation of course has a strong impact on the water quality in the river. Within the scope of the Master Plan, remedies of this situation are to be proposed.

5.2. PROPOSED REMEDIAL SCHEME

As described above, the rainfall induced infiltration represents in the Prague case quite an important issue which has to be analysed and the measures for its reduction must be proposed. With respect to infiltration, either point or diffuse sources can be distinguished. While point sources can be controlled and sewers reconstructed (pipe lining, etc.), there are hardly any possibilities to reduce the diffuse infiltration to the sewer system. The main concern of the presented work was therefore directed to identifying point sources of infiltration or the areas with high infiltration rates.

For that case, the key parameter of "sewer unit length infiltration" has been defined following Gustaffson [4]. This parameter represents a dimensionless infiltration volume from subcatchments. The general evaluation of the Prague sewer system was performed using this key parameter. The graphical representation of this evaluation is found in Figure 16.

It can be noted that the curvature of the infiltration graph indicates the global ratio of diffuse infiltration. The more linear is the graph, the more uniform infiltration is expected in the whole system, and fewer possibilities of infiltration reduction are available. High line gradients on the other hand indicate high unit infiltration in a particular subcatchment, which could be caused by point sources. In these cases, the infiltration can be evaluated in more detail, while searching for point infiltration sources. The identification of high infiltration subcatchments directs the project

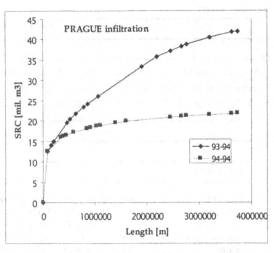

Figure 16. Unit length infiltration graph for the Prague Study

team to the indicated system weak points, where more precise measurements, modelling and finally TV inspection can take place.

In the Prague sewer system, the large infiltration rate is seen in the first graph line. This line represents for both simulated cases the contribution of subcatchment REST located in the vicinity of the WWTP. Also, the remaining part of the graph has a curved shape. Consequently, also catchments SBII, Mo and 112B indicate, along with the REST, the point infiltration sources. The infiltration volume from these catchments represents approximately 10% of the annually treated wastewater volume. The aforementioned catchments will be evaluated in more detail during the subsequent project phases, in which detailed evaluations of individual subcatchments will be performed. TV inspections will be proposed on the basis on obtained results and, consequently, a reconstruction of relevant sewer sections will be carried out.

6. Conclusions

The Urban Drainage Master Plan for the City of Prague represents a very large, methodologically and technologically complex project. The modelling of surface runoff processes covers only a portion of the whole work. The project team was confronted with similar questions and tasks in all modelling activities.

It can be concluded from the above described facts that the infiltration phenomenon plays in Prague urban drainage quite an important role. It affects both the treated volumes and CSO spills (volumes and peaks). Moreover, the infiltration represents a substantial portion of the annually treated wastewater volumes at WWTP and volumes spilled from the WWTP emergency overflow weir to the Vltava River. Along with that, the importance of a catchment "hydrological memory" with respect to the hydrological behaviour of the urban drainage system has been recognized during the project execution and has been taken into account in the master plan considerations.

Generally it can be concluded that the present methodology and technology for master planning still have some space for improvements and updates. The whole Prague project data management would have been rather tedious without some substantial additional programming work. However, the Prague project team is convinced that mathematical modelling of natural processes in urban areas can be currently successfully applied without any doubts even in large size projects as is the case of the Master Plan for the City of Prague.

7. References

1. Ball, J.E., Luk, K.C. (1996) Determination of the rainfall distribution over catchment using hydrodynamic tools, *Hydroinformatics 96*, Balkema, Rotterdam, ISBN 90 54 10 852 5.
2. Gustaffson, L.G., Lindberg, S., Olsson, R. (1991) Modelling of the indirect runoff component in urban areas, in C. Maksimovic (ed.), *Proceedings from International conference on urban drainage models and new technologies*, Dubrovnik, June 17-21, 1991, pp. 127-133.
3. Metelka T., Mucha A., Zeman E., Kuby R., Gustafsson L.G., Mark, O.(1998) Urban Drainage Master Planning – long term behaviour analysis, *Hydroinformatics 98*, Balkema, Rotterdam, ISBN 90 5410 9831.
4. Gustafsson, A.M. (1994) Key Numbers for Infiltration/Inflow in Sewer Systems, in *7th European Junior Scientist Workshop on Integrated Urban Storm Runoff*, Cernice Castle - Bechyne - Czech Republic, 1994.

APPLICATION OF 3D COMPLEX MODELLING IN SIMULATION OF EXTREME DISCHARGES IN URBAN AREAS

Prediction and evaluation of effects with the aid of an integrated sewer/aquifer model

R. KUBÝ[1] & L-G. GUSTAFSSON[2]
[1]*DHI Hydroinform, Na vrších 5, 100 00 Praha 10, Czech Republic*
[2]*DHI Sverige, Honnorsgatan 16, Vaxjo, Sweden*

1. Introduction

The City of Ceske Budejovice has initiated a flood model project of the Vltava and Malse Rivers including the catchment of Dobrovodsky creek in 1998. The aim of the project was to create a mathematical model for the reconstruction of historical flood events and the evaluation of impacts of possible technical solutions that could be introduced in order to prevent the floods in the future. The model should describe both the hydrological phenomena and the hydraulic capacity. As the catchment is partly urbanised the model should also include a model of the sewer network.

J. Marsalek et al. (eds.), Advances in Urban Stormwater and Agricultural Runoff Source Controls, 221–231.

2.   Project area

The catchment of Dobrovodsky creek is in the northeastern part of Ceské Budejovice. The total catchment area is 21 km². The upper part of the catchment has a mild slope, the central part has a steep slope and the downstream part is flat. The downstream part is partly urbanised, and includes an area set for a new development. Floods have a

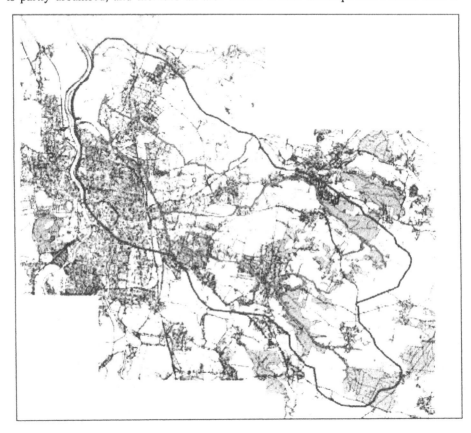

Figure 1. The project area

negative impact on the downstream catchment part. The historical urbanisation has changed the route of Dobrovodsky creek and consequently decreased its longitudinal slope. There are sewer collectors with several combined sewer overflows (CSOs) along the creek. The difference in elevations of the creek and sewer network is very small and the mutual interaction between these two systems is quite frequent.

3. Monitoring

Although the historical flood event in 1991 was well documented, there were insufficient input data for the calibration and verification of the proposed models. A short-term, comprehensive monitoring campaign was therefore executed within the frame of the project. The water levels in wells as well as water levels in creeks, the discharges into the creeks and into the sewers, and rain intensities were monitored for several months. The results of the monitoring provided the ground for analysis of the hydrological and hydraulic phenomena and also for the calibration of the models. The performance of the calibrated model was then verified using historical observations.

4. Mathematical modelling

The creation of the model of the catchment was done in three stages. Firstly the hydrological model of the catchment for the simulation of rainfall/runoff processes and groundwater flow was carried out using the MIKE SHE software. Then the model of the creek, for the description of hydraulic phenomena and analysis of hydraulic capacities, was created on the basis of MIKE 11 and a hydrodynamic model of the sewer network was prepared in the MOUSE software. Finally a flood model based on MIKE 11 1D+ was prepared for the downstream part of the catchment.

4.1. MODELLING TOOLS USED

4.1.1. MIKE SHE
MIKE SHE is a deterministic, widely used, physically based modelling system for simulation of hydrological processes in the land phase of the hydrological cycle. The model is applicable to a wide range of water resources and environmental problems related to surface water and groundwater systems and to the dynamic interaction between these two systems. The modelling package comprises a number of pre- and postprocessors to facilitate the input of data and the analysis of simulated results. The package includes, for example, space interpolation routines, graphical editing facilities, plots of variations in space and time of any variable, number of animation tools, etc.[1]

MIKE SHE is well capable of simulating the variations in hydraulic heads, flows and water storage on the catchment surface, in rivers and in the unsaturated and saturated subsurface zones. A network of grid squares represents the spatial variation of meteorological input data and catchment characteristics. Within each grid square the soil profile is divided into a series of vertical layers. The model structure is illustrated in Figure 2.

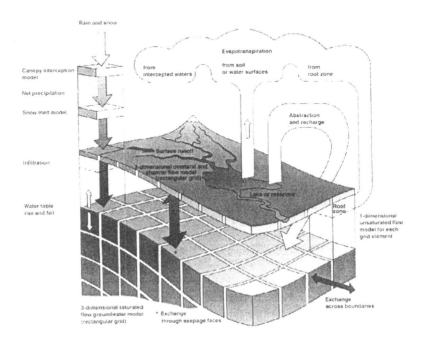

Figure 2. The structure of the MIKE SHE model

In addition to the water movement module, the modelling system includes modules for spreading and degradation of pollutants, mapping of capture zones, wetland ecology, etc. For more complex river hydraulics, the river-modelling package MIKE11 can be used and be fully integrated with the MIKE SHE software.

4.1.2. MOUSE
MOUSE is a well known, modelling package for urban drainage and sewer systems [2]. In addition to the fully hydrodynamic pipe flow model and conceptual surface runoff and inflow/infiltration modelling capabilities, the system includes modules for modelling pollutant transport and the transport of sediments, as well as off-line and on-line modelling of real time control.

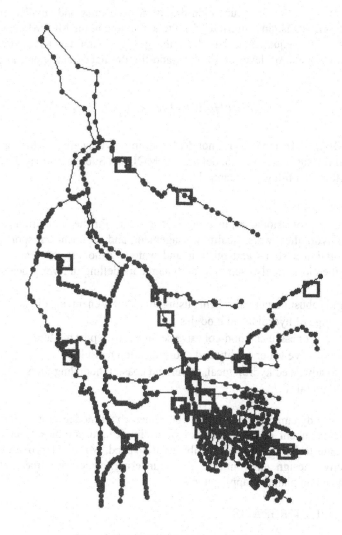

Figure 3. The model of the sewer network

The MOUSE and MIKE SHE models are coupled through the following three physical elements:

- The reciprocal interaction between pipes and the aquifer[3]. The interaction equation is based on the temporal and spatial variation of the water pressure in the pipes and the surrounding aquifer, the pipe surface exposed to the ground water (or to waste water, if exfiltration occurs), and a leakage coefficient. The equation reads:

$$Qin = (Hgw - Hsewer) \times Psewer \times C \tag{1}$$

- A one-way flow from foundation drains, service lines and smaller pipe systems (not described as pipes in MOUSE) into a manhole in the MOUSE sewer network. The drainage equation is based on the ground water level, the drainage level (typical foundation level or similar), and the drainage coefficient, and expressed as

$$Qdr = (Hgw - Hdr) \times Agrid \times Cdr \qquad (2)$$

- The MIKE SHE model does not deal with impervious areas, which are described within the surface runoff model used in MOUSE. Such areas may amount to only a fraction of a full grid square[1].

4.1.3. MIKE 11

MIKE 11 is a professional engineering software tool for the simulation of hydrologic processes, hydraulics, water quality management, and sediment transport in estuaries, rivers, irrigation systems and other inland waters. Amongst the features that make MIKE 11 the most widely used tool for dynamic modelling of river processes are:

- fast and robust numerical scheme using the Saint-Venant equations
- wide range of hydrological modules
- advanced cohesive and non-cohesive sediment transport modules
- comprehensive water quality and eutrophication modules
- links to advanced hydrological, sewer and coastal modelling tools
- add-in modules

MIKE 11 is a dynamic, user-friendly, one-dimensional modelling tool for the detailed design, management and operation of both simple and complex river and channel systems. Due to its exceptional flexibility and speed, MIKE 11 provides a complete and effective design environment for engineering, water resources, water quality management and planning applications.

4.2. MODEL INTEGRATION

Each of the models and appropriate software tools plays its role within the project. The hydrological model of MIKE 11 produces the flow data for the river model of MIKE 11 and/or flood model of MIKE 11 1D+ that simulates the flood wave. The sewer model in MOUSE simulates either the flow from a CSO into a creek and/or the inlet of water from a creek into the sewer through a CSO in case of high water level in the creek. While a two-way interaction may take place according to the flow conditions in the creek and in the sewer, the drainage of water from the aquifer into the sewer pipes is neglected because it has no significant impact on the flood flows. The models were integrated off line; i.e. the resulting time series from one model was used as boundary conditions in another model.

5. Application of the models in the catchment

5.1. PREPARATION OF THE MODELS

In the model, the catchment area is represented by several layers of 200 m x 200 m grid describing the surface of the catchment, the unsaturated zone, and the saturated zone. The input data needed for the model preparation are:
- The topological data
- The meteorological and hydrological data
- The topographical data and hydraulic data for creeks
- The soil parameters
- The land use and crop characteristics
- The impervious surface and runoff characteristics
- The geological information

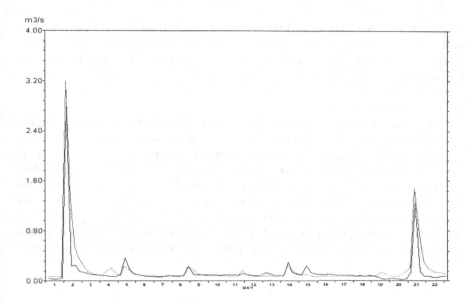

Figure 4. Example of creek model calibration at cross-section P3
(full line = observations, dotted line = simulations with MIKE SHE)

Although the goal of the project was to study high flows during rainy periods, the model is also ready for analysis of low flows during dry weather conditions. The model in MIKE SHE was calibrated by using the data obtained from the monitoring campaign mentioned earlier. The campaign was divided into two parts. The first part took place from June 1998 to October 1998, and the second part was realised from March 1999 to May 1999. The models were calibrated on the following two sets of values:

- The flows in creeks and sewers
- The water levels in wells.

5.2. THE VERIFICATION OF THE MODELS USING THE DATA FROM THE FLOOD IN 1991

There was a heavy rain storm on July 27, 1991, in Ceske Budejovice that caused a fast build up of flow, reaching the peak values in the evening of the same day. The total measured rain depth was 78 mm, out of which 58 mm fell during one hour. This rain event caused a flood with a return period between 50 and 100 years.

6. Evaluation of proposed technical solutions with the aid of models

6.1. PROPOSED SOLUTIONS

Two main scenarios were studied for control of floods in the future. The first one is based on the construction of retention dams with sufficient volume, i.e. the "retention scenario". The second one is based on a change of the current land use, i.e. the "land use scenario".

6.1.1. Retention scenario

The dams could be constructed in the upper part of the catchment only, where the first one could be located on the Dobrovodsky creek and the second one on the Vratecky creek. The potential retention basins could control a discharge from approximately one third of the total catchment area. That would allow for only a limited flow management, and furthermore the capacity problems of the creek were observed in the downstream, urbanised part of the catchment. There are two main reasons for the low hydraulics capacity of the downstream parts of the creeks. The first one is low slope due to the changes of the topology of the creek. The second one is a number of bridges and other man-made structures that decrease the flow area of the cross-section profiles.

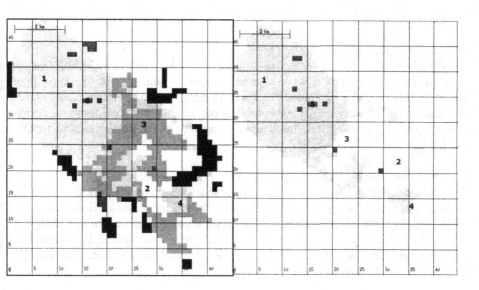

1-non cultivated meadows	1-non cultivated meadows
2-forests	2-forests
3-cultivated meadows	3-mixed forests
4-meadows	4-meadows
5-root and tuber crops	5-water bodies
6-water bodies	

Figure 5. The changes in land use

6.1.2. Land use scenario

Certain changes in the land use were suggested, as shown in Figure 5. The changes were expected to influence the flows in the creeks within the catchment. The models were used to predict these changes.

Flood control through "optimal" land use has been a frequently discussed topic in the Czech republic in recent years. Ecological organisations often argue that the changes of land use, such as the revitalisation of an area, will protect urbanised areas against large floods.

The runoff from urbanised areas is faster then the runoff from rural areas. In other words, the time of concentration of an urbanised part of the catchment is shorter than the time of concentration of the rural areas of the catchment (assuming there are no large differences in slope). This fact turned out to be important when evaluating this proposed solution.

230

6.2. EVALUATION

6.2.1. Evaluation of the "retention scenario"

Results obtained from the simulation of the retention scenario indicated that the retention dams would enable us to control, at least to some extent, the large flood events. Because of their location within the catchment, they could not have much effect on smaller floods, especially in the downstream part of the creek. To achieve the desired protection effects of the dams, the hydraulic capacity of the downstream part of the creek would have to be improved.

6.2.2. Evaluation of the "land use scenario"

The models demonstrated that the influence of this scenario on the peak flows of large flood events is relatively small, but it is possible to influence smaller floods and flows under the dry weather conditions. The analysis showed that the planned change of urbanisation would have a significant impact on the total flow volume, but it would not influence much the peak flows. The change of urbanisation would increase the local flow maximum, but not the total flow maximum.

7. Conclusions

The set of mathematical models, i.e., MIKE SHE hydrological model, MIKE 11 river model and MOUSE sewer network model, was applied to the catchment of the Dobrovodsky creek to evaluate the "retention scenario" and the "land use scenario". These were the two technical solutions proposed to mitigate floods in the downstream, urbanised part of the creek in the future.

It was found by applying models to the retention scenario, that the dams could influence the large floods, assuming that the hydraulic capacity of the creek would be changed.

The models demonstrated that land use scenario would have a low impact on large floods, but could affect small floods and/or dry weather flow condition.

The low hydraulic capacity of the creek channel was found to be the most adverse feature for the flood control in the future. The study therefore proposed to direct the investment to improve the hydraulic capacity of the creek in the urbanised part of the catchment. The retention dams in the upstream part of the catchment could be applied as an additional technical measure to improve the level of protection of the urbanised area. The planned urbanisation would have a low negative impact on the floods, mainly because the time of concentration of the urbanised part of the catchment is significantly shorter that the time of concentration of the rural areas.

Besides these technical solutions, the study also suggested to make a political decision about the desired level of protection against the floods.

8. References

1. Abbott, M. (1986) An Introduction to the European Hydrological System, Systeme Hydrologique Europeen "SHE". Journal of Hydrology **87**,45-59,61-77.
2. Lindberg, S. et al (1989) An integrated PC-Modelling System for Hydraulic Analysis of Drainage Systems. The First Australian Conference on Technical Computing in the Water Industry: WATERCOMP '89, Melbourne, Australia.
3. Gustafsson, L-G. (2000) Alternative Drainage Schemes for Reduction of Inflow/Infiltration – Prediction and Follow-up of Effects with the Aid if an Integrated Sewer/Aquifer Model. International Conference on Urban Drainage via Internet, Prague, Czech Republic, 2000.

FLOOD AND SOURCE CONTROL STUDIES IN SLOVAKIA

P. PETROVIC

Water Research Institute, Nab. gen. Svobodu 5, 812 49 Bratislava, Slovakia

1. Introduction

Slovakia is a country lying on the "roof" of Europe with runoff to the Baltic and Black seas. The outflow from about 98% of precipitation is through the Danube River. With regard to the Danube river, Slovakia has 22.4 km inside the country, 7.6 km are shared as a boundary with Austria and 142.1 km form the state boundary with Hungary.

From the point of view of flood protection, there are two different approaches to risk analysis. The volume and discharge of flood water that comes from upper parts of the Danube basin can be forecasted and/or modelled with the help of data from the upper regions' hydrological surveying services. The flood risk evaluation for the territory inside the country is a very complex and difficult problem due to its orographic and meteorological conditions. Precipitation amounts range from less than 450 mm a year (in the Danubian lowland) to more than 2500 mm a year (in the High Tatras region). The spring discharge caused by the snowmelt events is mostly estimated by using models, like the empirical regression model (ERM) [1]. Variability of rainfall events is very high; the observed maximum daily precipitation (locality Salka, 1957, July 12th) is 231.9 mm [2]. Such an amount of precipitation, especially in mountainous areas, can cause a serious flood wave, which can increase as a result of both a saturated soil layer (antecedent precipitation) and the areal extent of extreme precipitation.

A study of flood risk evaluation, early warning system and assessment of the flood protection measures is of the continuing interest of the water management organisations and respective administrative authorities.

2. Legislative and Administrative Background of Flood Management

The latest governmental decision dealing with the flood issue was agreed on 19 January 2000 under the sequence number 31/2000. The Government of Slovakia has agreed to a decision support system proposal for a "Programme of flood protection in Slovakia until 2010" with a step by step application of short, middle and long-term measures proposed in the Programme. This includes the expected financing method and a set of governmental recommendations for flood protection management based on the materials prepared by the Ministry of Agriculture of the Slovak Republic.

The financial base for the realisation of decision support proposals is at the level of $17.8 * 10^9$ SKK (approximately 410 million Euro) and additional support for

J. Marsalek et al. (eds.), Advances in Urban Stormwater and Agricultural Runoff Source Controls, 233–244.
© 2001 *Kluwer Academic Publishers. Printed in the Netherlands.*

research activities is expected at the level of $223.8 * 10^6$ SKK (5.2 million Euro). Investment and maintenance cost of research results application is not included in the proposed budget scheme.

The responsibility for meeting the Programme requirements is distributed among the Ministries of Agriculture, Finance, the Interior, Environment and Education. The main actions related to this governmental decision are as follows.

- Prepare a new version of governmental guidelines (No. 32/1975) for the flood events management and for financing the necessary management and rescue activities in time of flood events.
- Prepare a new version of "Water law" (138/1973 Zb.) in accord with the policy of the EC.
- Prepare a new version of the State Administration of the Water Management Law (No.135/1974 Zb.).
- Prepare guidelines for the evaluation, verification and (partial) reimbursement of flood damages.
- Prepare and introduce a set of short term measures for flood prevention and protection.
- Develop and set into operation an early warning system.
- Include the supporting Flood Research Programme Proposal into the State Plan of Research and Technical Development (Ministry of Education).

The research and technical development programme has to be oriented towards the following:
- hydrological and climatic flood conditions,
- rainfall – runoff process and design values for precipitation and flood events (risk scenario),
- landscape and geomorphologic structure of river basins, analysis of regions at risk and influencing factors,
- interaction of surface waters, soil and groundwater during flood events,
- flood development in the river network and the necessary technical protection measures, and
- the role of dams and reservoirs in flood protection.

This is the Slovak internal situation in the area of flood management. With respect to the European flood protection situation, it seems to be useful to consider the fact that the Danube River is the second longest river in Europe. In the area with countries at different levels of economic and political development, the International Commission for the Danube River Protection (ICDRP) was recently created. This is a new body and until now it has faced mostly water quality problems. A similar International Commission for the Protection of the Rhine has a commonly agreed Action Plan on Flood Defence [3]. Such a plan would be a serious step forward in the international co-operation in this part of Europe. Until now such a common programme proposal has not been considered by the ICDRP Board. We, in the Danube River basin, are facing more complicated topics. Some states belong to the EU, some are in the position of associated countries, some countries belong to the "TACIS" area and some countries, contributing to the Danube River runoff, seem to be outside of the above mentioned schemes (i.e. Yugoslavia, Albania, Macedonia). Under such varied conditions, it seems that a common approach to flood protection study is not realistic.

The International Hydrological Programme creates the only exception. Within the regional co-operation in the field of hydrology within the framework of UNESCO there is set of programmes oriented mostly to hydrological topics. These include a study about the coincidence of flood events [4]. More details about this co-operation can be found in the Proceedings of the XXth Danubian Conference on Hydrological Forecasting (Bratislava, September 2000, [5]).

In fact, Slovak hydrologists have been dealing with flood studies for a very long time. This paper does not include all the past or ongoing activities. It seems reasonable to mention, that the main responsibilities are shared by

- the Slovak Hydrometeorological Institute (controlled by the Ministry of Environment), which is responsible for monitoring water quantity and quality and for the (among others) flood forecasts and

- the Water Research Institute (controlled by the Ministry of Agriculture), which is responsible for "engineering water" in Slovakia, including flood risk evaluation, flood protection measures development and for water management methodology in practice.

There exist, of course, other institutions dealing with water problem studies. These include the Slovak Academy of Sciences, the universities and the Slovak Water Management Enterprise with four regional River Authorities Offices.

3. Some Projects Dealing with Flood Risk Assessment

In general, flood evaluation problems are highly related to the nature of the input data and the purpose of the analysis.

As mentioned above, flood protection, risk and management assessment can be divided into real-time questions and proposals for long-term policy measures. The real-time problems are closely related to the meteorological input, where the flood forecast should be based on the "as fast as possible" information about the forthcoming rainfall event. From this point of view, the use of radar input and follow-up action are of interest. The current situation with the weather radar usage in Slovakia is not satisfactory. There are two radars in operation; the one near Bratislava works well, but the older one on Kojsovska Hola is obsolete, and the middle region between these two radar fields (see Figure 1) is not covered at all.

236

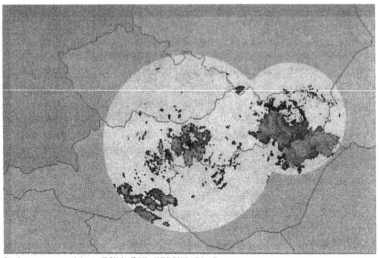

Radar composite: ECHO TOP HEIGHT [hm]

Maly Javornik   : 11-JUL-99 16:15   C   channel
Kojsevska hola   : 11-JUL-99 16:12   X&S channels' max.

0   20   40   60   80   100   120   140   160 [hm]

Figure 1. A radar image of the meteorological situation for Slovakia.

Information for the radar in Bratislava is available on the INTERNET [6] (reflectivity, cloud top elevation, tracing of thunderstorm paths, cumulative precipitation for the latest 1, 3, 6 and 24 hours).

For weather forecasts, it is necessary to have more information than presently available; other information can be obtained from the Czech radar and also from the INTERNET [7]. OPERA - the WMO - European project [8] dealing with the unified cumulative radar information is under development. It can produce Composite European RADar (CERAD) information on the digital radar echowave situation (see Figure 2) and accumulated rainfall. This system, which is in the experimental operation stage, is managed by the Austrian Central Institute for Meteorology and Geodynamik (ZAMG). The distributed information is at present not accessible from "outside", but access can be negotiated by weather forecasting services.

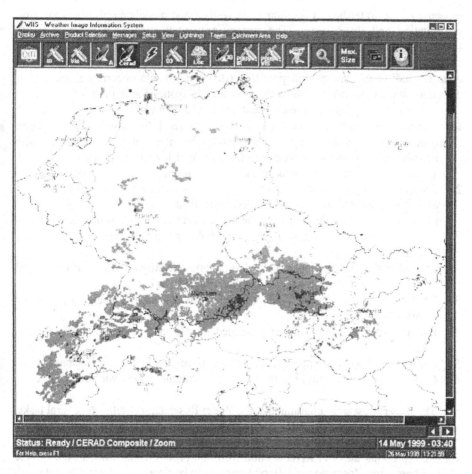

Figure 2. The composite European RADar image.

The network of radars in Europe is under development and, for locations in central Slovakia, there is a possibility to get Doppler-type radar information from Katowice, which is a new station and will be included in the system very soon. A similar project for the whole world is planned by the WMO [9].

The above-mentioned available information, combined with meteorological satellite images and synoptic meteorology methods, is sufficient to assess of the probable intensity and extent of rainfall events. This information can be included in the early warning system, which has to be developed within the realisation of tasks related to the decision of the Slovak Government from January 2000. This topic is within the responsibility mainly of the Slovak Hydrometeorological Institute and the Slovak Ministry of Environment.

Having information about a "given" rain event (neglecting the snowmelt impact), the question is what will happen in the specified river basin.

The problem solution strategy is based on mathematical modelling methods. The first step is determination of the actual extent and intensity of precipitation and the response of the watershed. It is common to verify flood models on historical extreme precipitation and flood events. The surface precipitation gauge network contains singular points (the mean density in Slovakia is about 1 station per 80 km^2). The existing statistical analysis of the maximum possible precipitation distribution was done for individual stations, independent of the time of rain occurrence, and hence does not give any information about the precipitation spatial distribution.

The main steps of flood risk evaluation are as follows.

1. Perform the depth-area-duration (DAD) analysis [10], which has not been common in Slovakia. It is possible to consider the impact of climate change on the rainfall intensity and duration due to the increased level of tropopause (results in an increased volume in the atmosphere for weather phenomena development).

2. Set up the rainfall – runoff model for the basin on existing flood events by
   a) using a lumped model for the whole watershed,
   b) using the vertical zoning in the model (clustered parameters approach) - evaluation for each zone or selected subbasins in the given watershed [11], and
   c) using the GIS approach, considering terrain configuration, land cover and slope (distributed parameters in space and perhaps in time).

3. Process expected design precipitation events (scenario) using the models mentioned above.

4. Evaluate the discharge volume generated by the chosen precipitation event(s) scenario.

5. Perform the flood wave propagation and transformation in the river channel.

6. Evaluate the flood extent from the flood wave volume in the river basin.

The present research programme at the Water Research Institute has several tasks, which cover at least a part of the above scheme. The studied areas can be seen in Figure 3; Case 1 is the Kysuca River, where a rainfall-runoff study is going to be done. Case 2 represents the Váh River section from the Hricov Dam down to the Nosice Dam, where a transformation of a flood wave has to be done and a hydraulic approach has been chosen. Both areas are in the Váh River basin.

Figure 3. Scheme of selected regions for flood risk assessment.

The Váh River is 402 km long from its source to the Danube River. The area of this large watershed is about 19,700 km² (34% of the Slovak Republic area) of which 10,247 km² is for the main course of the Váh River (excluding contributions of Nitra, Zitava, Dudvah Rivers). About 2.2 million inhabitants live in these watersheds. In addition to the above studies, several other tasks for particular parts of the Váh River are solved by different groups of our experts.

The main river has been intensively regulated for hydropower production, navigation and flood control purposes, for a long time. There is a cascade of 17 reservoirs greatly varying in volume: the first one (Ladce) was introduced in 1934, the biggest one (ORAVA-TVRDOSIN – the Orava River) was put in operation in 1952 and the last one (ZILINA – the Váh River) is going into operation now.

3.1. THE KYSUCA RIVER BASIN

The Kysuca River represents a tributary of the Váh River and lies in an area adjacent to the Moravian basin, which incurred major damage during the flood in 1997.

The total area of the subbasin is 1037 km² and the length of the main channel of the Kysuca River is 66 km. The highest point in the watershed is 1326 m a.s.l. (Velka

Raca) and the mouth of the Kysuca River (at the Váh River) has an elevation of 325 m a.s.l. In co-operation with the Slovak Hydrometeorological Institute (SHMU), the DAD analysis is in progress. For the areal distribution of precipitation and evaluation of some other terrain characteristics, the digital terrain model will be processed from the basic geographical map in a digital vector form (including contour lines) at a scale of 1 : 50 000. Part of the watershed is available in orthophotomaps at a scale of 1 : 10 000. Selected critical nodes can be evaluated from the paper maps at a scale of 1 : 10 000 using scanning maps. We do have an extension for the ArcView programme, which allows co-ordinates to be set to the grid image, so that the full supplementary information is available, if necessary.

The next step in the SHMU will assess the climate change impact on the extreme precipitation (both in space and time distribution) and develop some "design" scenarios for a probable critical situation.

The Water Research Institute (VUVH) will fine-tune the rainfall-runoff model. About 20 flood waves, measured since 1960 (since that time the hydrometeorological data have been stored in digital form at the SHMU) will be used for the model assessment and calibration. The "vertical-layered" version of ERM will be applied. The program is written in FORTRAN 77 and computation of the necessary data outside the ArcView GIS environment will have to be solved.

The last step will be the evaluation of new runoff phenomena using the model calibrated on historical flood events with the new design precipitation considering climate change scenario with a daily time step.

3.2. THE VÁH RIVER SECTION FROM HRICOV TO NOSICE

The river section in the middle part of the river basin (between Hricov and Nosice) is at present one of the high risk areas; a new highway has been constructed and there are some problems with flood management.

Thanks to the activities of the Institute of Informatics of the Slovak Academy of Sciences (II SAS), the flood risk evaluation and flood extent maps assessment was included in the 5th Framework Programme of the EC as a project ANFAS (datA fusioN for Flood Analysis and decision Support). The aim of the project is to use all the available data and multiprocessor techniques for flood risk and development assessments, including presentation of flood extend maps in a suitable GIS package. The project belongs to the IST section in the EC RTD Projects. The main co-ordination institution is ERCIM in France [12]. Participating organisations in Slovakia are II SAS, the Water Research Institute (VUVH), the Váh River Authority (VRA) office and, indirectly, the Slovak Hydrometeorological Institute.

The ANFAS project will

- use data from the most advanced acquisition technology, in particular remote sensing imagery (such as optical radar, and interferometry radar) and incorporate them into a conventional Geographical Information System database;
- develop advanced data processing tools for scenario modelling and flood simulation, including computer vision and scientific computing in order to take into account information extracted from real images for the calculation of parameters of the simulation model;

- have strong end-users involvement through definition of the objectives, concept of the System, evaluation of the simulation results, and by ensuring that the System answers to "terrain needs" and is well adapted to practical use.

The final selection of the river section was done by a visit of an international expert group to the watershed and the section from Hricov to Nosice was chosen as the pilot site (Figure 4).

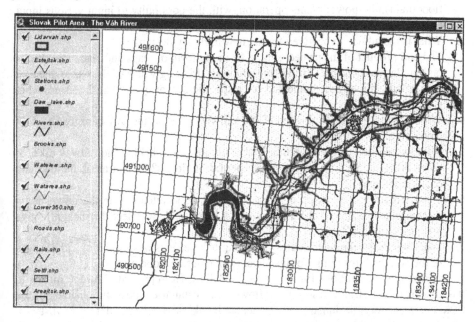

Figure 4. The selected pilot site – the Váh River from Hricov to Nosice.

The Váh River Authority office plays the role of the "end user" in this RTD project. Its main mission is concentrated on water utilisation as regulated by the relevant Slovak laws, flow control, allocation of water quantity and quality, and management and protection of surface and ground waters. One of the tasks in the project was formulation of the end user requirements, which has to be considered in the project preparation and operation. The requirements were prepared with respect to the present actual problems in the area, where the watercourse after the artificial bifurcation at the Hricov reservoirs has one branch as a natural river channel where in time of low flow only a minimum discharge is satisfied. The second branch is an artificial canal with two power plants, which are used for power generation - according to the available water - usually at times of peak energy consumption.

The section of the Váh River between the Hricov reservoir and the Nosice reservoir is of major interest for the Slovak end-users within the frame of the ANFAS project. First, flood damages may occur in the area in case of extreme floods. Second, an important highway under design will cross the Váh River and its floodplain several times. Third, it is strategic to issue flood warnings and control floods as far upstream as possible in the watershed (i.e., upstream of the Nosice reservoir) in order to ensure efficient protection in the lower basin which is much more populated.

After summarising the above information, further practical questions related to end user needs were formulated.

1. Hydraulic modelling should investigate the whole stretch from the Hricov reservoir to the Nosice reservoir, considering the natural Váh channel (the hydropower canal on the right bank is an areal boundary condition for the floodplain), the power plants operation, with the possibility to input various input hydrographs (flood scenarios) at Hricov and the Váh tributaries.
2. What is the advice to be given for the new highway design?
3. The end product of the project should be in the form of maps and statistical results on inundation damages. Are villages, houses, drinking water wells, waste disposal sites, roads, railways... inundated for a specific scenario, at what time step, what is the water level?
4. Which areas would be flooded if the railway dike on the left bank of the Váh River breaks?
5. What would be the water volume that could be stored behind the railway dike, or other dikes, depending on the return period of floods? Which areas include and exclude villages?
6. What would be the effects of such voluntary flood retention on the downstream discharge and water levels? How would it be possible to select the most effective area to be flooded?
7. What would be the intermediate and long-term effects of morphological changes of the riverbed, resulting, for instance, from extreme flood events?

Boundary conditions for hydraulic modelling are represented by the inflow to the area from the Hricov reservoir, inflows from tributaries, possible changes in the flow cross-sections (voluntary or accidental break of a dike) and the downstream condition at the end of the backwater curve in the Nosice reservoir. This represents a stretch of 37 km of the river with a floodplain of 0.5 to 2 km wide. From a mathematical point of view, the downstream boundary condition is the elevation of the Nosice weir for the case of the maximum possible water level. Inflow scenarios will be a combination of probable discharges at all the nearest upstream gauging stations for runoff return periods from 5 to 100 years.

The general information about the hydrological situation in the selected section [13] is given in Table 1. From Table 1, it is evident that the tributaries collect water from an area of 850 km^2, which represents approximately 10 % of the area upstream of the upper reservoir.

TABLE 1. Main characteristics of the upper and lower reservoirs.

	Hricov	Nosice
Watershed area (km^2)	7145.8	7996.6
Q average (m^3/s)	121.0	131.0
Q minimum (m^3/s)	22.0	23.8
Flood Q$_{100}$ (m^3/s)	2300	2100
River chainage (km)	245	205
Reservoir volume (Mm^3)	8.5	36.0

About one-quarter of the project has been completed. Data assembling and the selection of models envisaged for this application is in progress. From the point of view of the end user, there are some interesting points. For example, this is the first time that LIDAR data will be used in Slovakia.

Effective use of data fusion involves the following existing data:
- river and floodplain cross-sections with intervals from 150 to 500 m; water level surveys during low and high flows; discharge measurements;
- elevation maps with a 1 : 50000 scale along with the ArcView files corresponding to the area (ArcView files were forwarded to project partners for further processing);
- one (composite for a good weather situation for the whole Slovakia) satellite image (Landsat); and
- existing aerial photos from which the orthophoto map of the studied area with resolution 1 : 10 000 and contour lines will be prepared.

Discharge scenarios will be prepared in co-operation with the Slovak Hydrometeorological Institute.

The international expert group will perform the final interpretation of results. The system of data fusion, developed in the course of the project, will stay for our use after the project completion in 2002.

4. Conclusions and Summary

Slovakia is undergoing economic transformation and aspires to join the EU. From this point of view, its participation in the IST project of the 5th EU framework RTD programme is a new, very welcome experience.

The Slovak authorities considering the development of water management and the situation related to an increasing risk and magnitude of flood damages decided to launch the state flood protection programme described herein.

The expected results will

- be based, in the Kysuca River basin, on a rainfall-runoff model for flood evaluation as the first step; a depth - area - duration analysis of precipitation will be performed; new design precipitation and its spatial and temporal distribution under the impact of climate change will be developed and new applications of models should lead to new design flood values;
- use hydraulic modelling methods in the section of the Váh River from Hricov to Nosice;

- improve knowledge about the role of GIS and digital terrain models in flood evaluation problems;
- support further flood evaluation by the system of data fusion for decision support systems and water management measure proposals;
- support further flood evaluation studies by calibrated hydraulic models for given areas for preparing and processing a new set of flood scenarios;
- supply very detailed information for evaluation of the flood risk and some protection and/or remediation measures for selected Váh River sections;
- contribute to high computer technique usage by connecting more PCs into a parallel CPU coupling network, with the aim to increase the evaluation effectiveness and shorten the time of event evaluation and result presentation; and
- increase the experience and skills of participating experts.

5. References

1. Turcan, J. and Petrovic, P. (1992) Contribution of Czechoslovakia (Use of the ERM in flood forecast and simulation) to the: WMO - Operational Hydrology Report No. 38 - Simulated Real-time Intercomparison of Hydrological Models - WMO - Publ. No. 779, ISBN 92-63-10779-3, Geneva.
2. Horecká, V. and Valovic, Š. (1991) Atmosférické zrázky [Atmospheric precipitation]. In: Zborník prác SHMÚ zv. 33/I, Alfa, Bratislava, ISBN 80-5-00888-0, 107 – 145, (in Slovak).
3. http://www.iksr.org/icpr/11uk.htm
4. UNESCO. (2000) Coincidence of flood flow of the Danube River and its tributaries (Co-ordination Prof. Dr. Stevan Prohaska, Federal Republic Yugoslavia). A Hydrological Monograph – Follow-up volume IV. Regional co-operation of the Danube countries in the frame of the International Hydrological Programme of UNESCO. ISBN 80-968282-3-1. Published by VÚVH Bratislava, 1999 (March 2000). (in English only, extended resume in German and Russian).
5. Miklánek P., Stancík, A. and Petrovic, P. (2000) Regional co-operation of the Danube Countries in hydrology in the frame of the IHP UNESCO (information of the co-ordinating country). Abstract in: XXth Conference of the Danubian Countries on Hydrological Forecasting and Hydrological Bases of Water Management. Bratislava, Slovakia, 4-8 September 2000, printed by SHMÚ, p. 140, full text of contribution is on CD: "XX. Conference of the Danubian Countries on Hydrological Forecasting and Hydrological Bases of Water Management", ISBN 80-85755-09-2 as a file "cl140.pdf". Available on request by Email.
6. http://www.shmu.sk/radar/js/index.html
7. http://www.chmi.cz/meteo/rad/data/index.html
8. http://www.eumetnet.eu.org/contopera.html OPERA (Operational Programme for the Exchange of weather RAdar information)
9. http://www.wmo.ch/web/www/BAS/ISS-Conference/Sun.html: The Auto and Interactive Image Processing System for TV Weather Forecasting by Sun Songqing, Lu Zhihao, Zhang Ruiyi (Shanghai TV Weather Center, Shanghai 200030)
10. Chow, V.T. (1964) *Handbook of Applied Hydrology*, Chapter 9 Rainfall by Ch. S. Gilman, McGraw-Hill, Inc., New York. pp. 9-32.
11. Turcan, J., Petrovic, P., and Ursíny P. (1985) Snowmelt runoff modelling using satellite information, Paper for the Round Table Discussion on Earth Remote Sensing and its Application in Hydrology and Water Management. Organised by Slovak Hydrometeorological Institute under the sponsorship of WMO in Kocovce.
12. http://www.ercim.org/ANFAS/description.html
13. Abaffy D. and Lukac M. (1991) *Dams and Reservoirs in Slovakia*, Alfa Bratislava, ISBN 80-05-00926-7 (in Slovak with an English summary).

GENETIC ALGORITHM TECHNIQUES FOR STORMWATER RUNOFF SOURCE CONTROL PLANNING AND DESIGN

E.G. BOURNASKI
Bulgarian Academy of Sciences, Institute of Water Problems,
Acad.G.Bontchev Street, Block 1, BG-1113 Sofia, BULGARIA

1. Introduction

Management of surface runoff and protection of water and environment, including source controls, are key elements of any catchment-wide development. The necessity of analysing the catchment area, the urban drainage system, and the receiving water as one single system is obvious. The related optimal planning and design of urban drainage networks belongs to the class of large combinatorial optimisation problems which are very difficult to handle using conventional operation research techniques. Introduced relatively recently to engineering, Genetic Algorithms (GAs) form a radically different approach to optimisation, which is refreshingly simple and ideally suited to the design of large complex systems. They are multiple point stochastic search techniques that simulate the natural process of evolution in a search for the optimal solution to a problem.

The aim of this paper is to briefly describe this technique and to present its possible application to optimisation of source control planning and design with respect to stormwater runoff. A potential application to an urban area in the Mesta (Nestos) river basin located on the border of Bulgaria and Greece is shortly outlined. Examples from other areas of water engineering, such as an optimal layout of branched pipe networks and optimal pump scheduling, are also provided.

GAs in water engineering should not be considered as a decision-making tool, but as a tool able to provide alternative solutions from which designers/decision makers may choose.

2. The Genetic Algorithm

The GA is probably the best known type of Evolution Programs (EPs), which are general search methods based on an artificial analogy with the principles of biological evolution[1],[2]. The analogy with nature is established by creating (within a computer) an initial population of individuals, and forming offsprings by combining the genetic information of parents and preferential survival of the fittest generation. These algorithms are best suited to solving combinatorial optimisation problems that cannot be solved using more conventional optimisation techniques, such as linear programming. Thus, they are often applied to large, complex problems that are non-linear and include multiple local optima.

J. Marsalek et al. (eds.), Advances in Urban Stormwater and Agricultural Runoff Source Controls, 245–254.
© 2001 *Kluwer Academic Publishers. Printed in the Netherlands.*

For example in GA, if the optimal design for a scheme is required, a population of different designs is considered. Each design is defined by numerical values (strings) for each of the design parameters and forms an individual (organism). Each organism is encoded as a chromosome by analogy with nature. Standard GAs use a binary alphabet to form chromosomes, although non-binary alphabets can be also used. The string can be of any length, and can represent actual discrete choices, or discrete values of a bounded continuous variable. An 8 bit (gene) string, Figure 1, represents for example up to $2^8 = 256$ discrete values for a design variable. The complete design can then be represented by a single long binary code consisting of all strings placed end to end, and is equivalent to a chromosome in the nature.

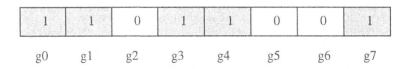

Figure 1. An 8-bit binary chromosome.

GAs are also referred to as stochastic optimisation techniques, because the solution space is searched by maintenance of population with diverse members, inheritance of genetic information from parents and by generating candidate solutions with the aid of a pseudo-random number generator.

3. GAs in Water Engineering and Management

After reported success in many domains [1], water engineering and management has started to benefit from the use of GAs. Cembrowicz [3] used GAs for designing the layout of a water distribution system in 1977. Walters and Savic [4] applied GAs for similar problem in the areas of water and gas supply and sewerage. Rauch and Harremoës [5] have demonstrated the GA's ability to assist modellers in producing a good solution to the problems of real-time control and runoff model calibration. Multiobjective GA techniques have been used to solve a groundwater pollution containment problem; hydraulic groundwater simulation models have been used in conjunction with GAs for respective management problems as reported in [6], where some results of stochastic optimisation techniques in hydraulic engineering and management are analysed.

Table 1 summarises some of the popular literature GA optimisation applications and shows the respective objective functions and design parameters.

TABLE 1. GA optimisation in hydraulic engineering and management, typical applications.

Problem	Objective function	Design parameters, Decision variables	Literature
Optimal layout selection of distribution network	minimum cost of a tree pipe network	pipe layout	[3, 4, 6]
Optimal pipe sizing for distribution network	minimum cost as a function of pipe diameters and lengths	pipe diameters	[7]
pump scheduling	minimisation of energy costs and pump switches	pump combinations in time intervals	[6, 8]
pressure regulation in pipe network	minimum pressure above the minimum required	setting isolating valves	[9]
optimisation of choice of pumps along a pipeline	minimum power requirement	pump combinations	[10]
calibration of water network model	minimum difference of field and model values of flow & pressure	pipe wall roughness for each pipe	[11]
least-cost design of water distribution network	minimum of cost function of pipe diameters and lengths	example of 8 pipe diameters	[12, 13]
water network rehabilitation	benefit of optimal invest. of some /all of a limited amount of money	replacement of pipes	[14, 15]
urban drainage system optimisation	minimal total cost to meet the specified hydraulic objectives	storage and flow attenuation structures	[16]
runoff model calibration	minimum difference of field and model values	roughness	[5]

4. GA Process for Source Control Planning and Design Optimisation of an Urban Drainage Project

A variety of GA applications has been presented since they were first introduced. In this paper, the approach of the University of Exeter Centre for Water Systems is used [16]. The basic steps of the relevant optimisation GA process for a combined sewer network include the 11 points described below.

4.1 DEFINE THE OBJECTIVE FUNCTION AND SOLUTION SEARCH SPACE

The objectives of any source control planning and design of sewer network project must be clearly identified prior to any design work. Although each design details are individual to the system, there is some commonality in overall objectives. Basic design objectives include reducing flooding (compared to the historical flood records, if available) and reducing the number and volume of spills to receiving watercourses, particularly bathing and recreational waters (related to receiving water quality and environmental protection). Some of these objectives are formalised in industry guidelines as minimum pipe cover levels, pipe gradients, velocities, flood alleviation levels, etc. [16]. The necessity of integrated approach in defining the objectives is obvious, by treating the urban catchment, the urban drainage system, and the receiving water as one single system [17].

The project costs need to be considered throughout the design process. There have been successful attempts to integrate the design network with several cost models [16], which provide the designer with information not only about the performance of the scheme but also outline implementation costs. This allows the designer to see the performance and cost of the design as it develops.

Thus the objective is generally to minimise the total cost to meet the specified hydraulic and environmental objectives, i.e., to reach a technical, ecological and economic optimum.

4.2 ESTABLISH A CHROMOSOME REPRESENTATION

As mentioned in Section 2, an appropriate format (a binary string) needs to be defined to permit the representation of the decision variables (the design options). For example an n-parameter solution (decision variable) $x = (x_1, x_2, ...x_n)$ may be represented as an $4 \times n$-bit binary chromosome (i.e. 4 bit per parameter, $x_1 = 0101$, $x_2 = 1001$, etc.).

4.3 GENERATE AN INITIAL 'POPULATION'

A number of organisms are randomly created each representing an individual system design. Some of the organisms may not be feasible. The initial population of perhaps 100 solutions is then allowed to evolve over a number of generations.

4.4 DECODE ORGANISMS TO SYSTEM NETWORKS

Each organism's chromosome is decoded into a representation of the system network model (a candidate solution). This is achieved by simply transforming binary strings into system parameter values.

4.5 ANALYSE SEWER NETWORK HYDRAULICS AND RECEIVING WATER IMPACTS

Operation of each network model is simulated hydraulically and the results of simulations are analysed. This means that the GA is integrated with a reliable computational model of hydraulic transport and water quality processes in the sewer as well as with a model for estimation of transport and water quality processes in the receiving water body (at least the stormwater runoff impacts). The aim is also to overcome the possible pollution of receiving waters by combined sewer overflows, which is the main weak point in most cases.

4.6 COMPUTE PENALTY COSTS

A graduated penalty cost is assigned to each solution as a function of the magnitude of its deficiency in meeting the specified hydraulic and environmental objectives. In complicated studies, such as the problem of source control planning and design of urban drainage systems, the environmental penalty is rather difficult to impose and its determination requires some experience.

4.7 COMPUTE TOTAL COST AND FITNESS

The total cost is computed for each solution as the sum of the network capital costs plus the penalty costs. The fittest solutions are those which exhibit good hydraulic and environmental performance and maintain low costs.

4.8 SELECT SOLUTIONS TO PARENT THE NEXT GENERATION

A new set of solutions is generated from the current population based on probability. The fittest solutions are given the highest probability of being selected; the least fit solutions may not survive and risk a permanent extinction. As a result, the new generation has on average a higher fitness than the old population. This selection process parallels nature's processes of 'survival of the fittest'.

4.9 APPLY CROSSOVER (RECOMBINATION) OPERATOR

The organisms (parents) selected in the previous step are grouped into partners and mated. Random genes of one of the partners' chromosome are swapped with the corresponding genes of the other partner to produce two offspring chromosomes representing two new organisms. Simply put, the binary strings (chromosomes) of both parents are broken at a randomly selected location and the second sections of the strings are switched to form two new solutions.

4.10 APPLY THE MUTATION OPERATOR

With a very low probability, randomly chosen genes in some chromosomes are altered (from 0 to 1 or vice-versa), e.g., changing one pipe size to another size, as a means of introducing or preserving valuable genetic material, which might otherwise be lost. Mutation does not have a dominant role in the process of evolution. If its probability is set too high, the search degenerates into a random process. As a simulation of a genetic process, a GA uses stochastic mechanisms, but the result is distinctly better than random.

4.11 REPEAT EVALUATION OF ORGANISMS AND PRODUCTION OF SUCCESSIVE GENERATIONS

Steps 4 through 10 are repeated thousands times until the termination objective is reached. At each repetition (generation), the best ever solution is stored so that the best solution is not lost in the next generation.

5. Some GAs Optimisation Examples

5.1 OPTIMAL LAYOUT SELECTION FOR BRANCHED PIPE NETWORKS

A simple GA was applied by Walter and Savic [4, 6] to study an optimal layout problem for water, gas or sewer pipe networks, with the objective to select the minimum cost set that is necessary to supply all given demands. For a base graph in which there are at most two choices per node (triple node), the design can be mapped exactly onto a binary string, as shown in Figure 1. However, for base graphs with more than two choices, an integer coding proves very effective.

5.1.1   Starting From a Directed Base Graph
The optimisation problem was defined as a search for the best layout starting from a directed base graph of possible pipeline connections. Any binary string represents a feasible layout, so crossover and mutation can be fully implemented without generation of non-feasible solutions. For a tree network, each node is supplied by just one pipe, directions of flow are fixed, and a layout can be defined by coding the choice of supply pipe for those nodes at which a choice exists. Although useful for sewer network systems, in which directions are largely predetermined, the above model is of a limited use for distribution networks.

5.1.2   Starting From a Non-directed Base Graph
Undirected base-graph networks are not only larger problems to solve, but also pose more difficulties than directed networks. A similar basic GA technique can be employed, however the crucial steps are in pooling the information from two parents and the formation of children. Figure 2 a shows a part of an undirected base-graph loop network used as an initial condition (population) for selection of a tree network. After some new populations, Figures 2 b and c show two parents (tree networks), and after their recombination a new graph containing some loops and a mixture of directed and undirected pipes is formed (Figure 2 d). This graph is then randomly 'mutated' by addition of a directed pipe (Figure 2 e). The children are 'grown' by a random selection and at the end of the whole process the resulting tree network is shown in Figure 2 f.

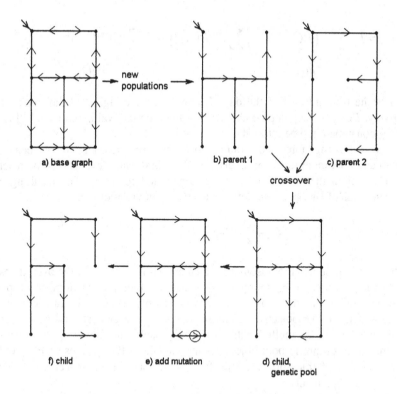

a) base graph new populations b) parent 1 crossover c) parent 2

f) child e) add mutation d) child, genetic pool

Figure 2. Some GA populations evolved in the optimal pipe layout problem.

5.2 OPTIMAL PUMP SCHEDULING IN WATER SUPPLY

Remarkable reductions in operating costs can be achieved by optimising the pump scheduling (PS) problem. Recent UK estimates of savings for an average system amount to ten percent of the current pumping cost; the overall annual energy cost for pumping in water supply is about £ 70 Million [8]. Classical optimisation methods may become inadequate when there are more than two reservoirs in the system or even for one-reservoir systems which have several different pump combinations or complicated system constraints. GAs treat discrete values used in pump scheduling models naturally and thus they seem well suited to this kind of optimisation.

For a system consisting of one water distribution reservoir which is supplied by four pumps (with different capacities) through a single water main [8], the optimisation period can be set to one day. To establish a chromosome representation, each pump during a certain time period (say 1 hour) is represented by one bit of the string (1 or 0 for switched on or off, respectively). A string consisting of $24 \times 4 = 96$ bits describes the problem completely. Thus, the overall number of possible solutions to the PS problem is 2^{96}. The electricity cost C_e is given as

$$C_e = T_n \sum_{i=1}^{7} C[PC(i)] + T_d \sum_{i=8}^{20} C[PC(i)] + T_n \sum_{i=21}^{24} C[PC(i)] \qquad (1)$$

night   day   night

where i is the time interval of the day, T_n and T_d is the night and day time tariff, respectively, $PC(i)$ is the pump combination of the time interval i, and $C[PC(i)]$ is the electricity consumed in time interval i.

The following three constraint violations are incorporated as penalties: minimum and maximum reservoir levels, and the initial water level should be reached in the reservoir at the end of the optimisation period reservoir. The resulting cost function in terms of energy consumption and constraint violations is:

$$C_{Cost} = C_e + \sum Penalty \qquad (2)$$

The fitness of individual chromosomes is calculated as the inverse of the overall cost.

Besides energy costs, pump maintenance by limiting the number of pump switches is another important cost factor in water supply. Thus in contrast to a single-criterion model, the two objectives of minimisation of energy costs and minimisation of pump switches form a multiobjective (MO) approach. The fitness of the organisms is determined according to both objectives individually. MO methods of PS problem provide a choice of trade-off solutions from which a decision maker can select a suitable one to implement.

There are several possibilities for improvement of GAs, including progressive assignment of penalties for constraint violation, introduction of feasibility of solutions as an additional objective, etc. The best solution identified by the GA shows, as expected, that as much of water as possible should be pumped using the cheap night time tariff. The reservoir is consequently as full as possible at the end of the cheap period and as empty as allowed at the end of the expensive period.

6. A Short Analysis of the Mesta River Basin as a Potential Area for GA Applications

The Mesta (Nestos) River basin is situated on the Bulgarian-Greek border. The river flows from the mountain Rilla and after a distance of about 125 km in Bulgaria and 140 km in Greece, it discharges into the Thracian (Aegean) Sea. With a mean altitude 1318 m the basin is found to be the highest river valley in Bulgaria (and may be on the Balkan Peninsula). Most of the catchment area is natural and rich in flora and fauna. However, state industrial enterprises established during the totalitarian regime as well as the forestry industry and agricultural activity caused disturbances of the natural river flow and deteriorated the water quality.

In Greece, 40% of the river basin contains one of the most important forests in Greece, unique in the world for its great variety of flora and fauna. Three dams are already being constructed there for water-power production and irrigation, and one more is planned. All this has positive and negative consequences for the environment.

However, apart from the changes due to the human activities, the valley and especially Mesta's Delta is still one of the most important wetlands in the country and is protected today by the Treaty of RAMSAR.

After political changes in 1989, the Bulgarian foreign policy supports joining the European Union and Euro-Atlantic structures. Prior to admission, many steps have to be taken. In the field of infrastructure and the water sector, huge investments will be needed, especially for the transboundary river catchment of the Mesta River. Existing disparities between national monitoring systems in the two countries are obstacles for integrated planning and control of water resources in the region. There is a need for tools to manage water resources in the most efficient way, predict their development, and avoid possible conflicting situations.

The towns on the Bulgarian territory (some with developed industry) are located on both sides of the Mesta River. The sewer systems are of the combined type and their design is based on the rational method applied during the totalitarian regime, without taking into account the impact on receiving waters (the Mesta River). There is a huge need for a new urban drainage master plan of this territory. Its methodological principle should be based on an integrated approach to rainfall-runoff, transport and treatment processes in the catchment, sewers, treatment plant and receiving water. The GA approach for source control planning and design, with respect to stormwater runoff, is the right optimisation tool that would help the designer and decision makers. It can contribute to the development of the infrastructure needed for further Greek-Bulgarian cooperation and joint water resources management in this transboundary river basin.

7. Conclusions

The GA approach described here for source control planning and design optimisation of stormwater runoff management is in its infancy, but there are many signs of its ability to assist the engineer in deriving a good solution to similar complex problems. It is important that the GA and the simulation modules of transport and water quality processes in the sewer network as well as in the receiving water are fully integrated. In fact, there may not be a single optimum solution, but instead a series of quasi-optimal solutions of this problem. The GA indeed may offer a range of solutions close to or on the optimal trade-off plane. It frees the engineer from making complex model alterations to find a single feasible design.

Experience has shown that often the best solution to a problem is not the solution that engineering judgement would initially suggest and a small change to a component may have significant effects. To analyse all these possibilities manually would be impossible. GA has the advantage of rapidly analysing many potential solutions and retain useful information in the population. It can also find key outlying components quickly and can incorporate them into the solution. Studies in water engineering modelling indicate that there is likely to be a cost saving on the GA based design.

One of the many catchment areas for potential application of this approach is the Mesta (Nestos) River basin situated on Bulgarian and Greek territories. This transboundary river exhibits favourable natural conditions and is the most important water source of the whole area. The GA optimisation technique for modern urban

254

drainage plans in this territory can lead towards improved drainage systems, requiring minimum investment, and causing less serious damage to the environment, thus contributing to future harmonisation in the field of environmental protection and joint development.

8. References

1. Goldberg D.E. (1989) *Genetic Algorithms in Search, Optimization and Machine Learning*, Addison-Wesley.
2. Michalewicz Z. (1996) *Genetic Algorithms + Data Structures = Evolution Programs.* 3rd Revised and extended Edition. Springer-Verlag, Berlin, Heidelberg, Germany.
3. Cembrowicz R.G. (1992) Water Supply Systems Optimization for Developing Countries, in Coulbeck B. and Evans E. (eds.), *Pipeline Systems*, Kluwer Academic Publishers, pp. 59-76.
4. Walters G. and Savic D. (1994) Optimal Design of Water Systems Using Genetic Algorithms and other Evolution Programs, in *Hydrosoft 94*, Blain W. and Ktsifarakis K. (eds.), *Vol.1, Water Resources and Distribution,* pp.19-26.
5. Rauch W., Harremoës P. (1998) On the Application of Evolution Programs in Urban Drainage Modelling. *4th International Conference on Developments in Urban Drainage Modelling*, 21-24 September 1998, London, UK.
6. Savic D., Walters G., (1996) Stochastic Optimization Techniques in Hydraulic Engineering and Management., *7th Int.Symposium on Stochastic Hydraulics*, Mackay, Australia, 1996.
7. Savic D., Walters G. (1995) Genetic Operators and Constraint Handling for Pipe Network Optimization (personal communication).
8. Savic D., Walters G., and Schwab M. (1997) Multiobjective Genetic Algorithms for Pump Scheduling in Water Supply, in *Workshop on Evolutionary Computing*, AISB '97, Manchester, UK.
9. Savic D., Walters G. (1995) Integration of a Model for Hydraulic Analysis of Water Distribution Networks with an Evolution Program for Pressure Regulation., *Microcomputers in Civil Engineering* 10, 219-229.
10. Goldberg D.E., Kuo C.H. (1987) Genetic Algorithms in Pipeline Optimization. *J. of Computing in Civil Engineering*, 1, 128-141.
11. Walters G.A, Savic D.A, de Schaetzen W, Atkinson R.M., Morley M. (1998) Calibration of Water Distribution Network Models Using Genetic Algorithms. *7th International Conference on Hydraulic Engineering Software*. Como, Italy. September 1998.
12. Savic D, Walters G. (1997) Genetic Algorithms for Least-Cost Design of Water Distribution Network, *Journal of Water Resources Planning and Management*, ASCE, 123, 67-77.
13. Atkinson R, Morley M, Walters G, Savic D. (1998) GANET: The Integration of GIS, Network Analysis and Genetic algorithm Optimization Software for Water Network Analysis, in *HydroInformatics 1998*, Babovic,V., Larsen,L.(Eds.) pp 347-362.
14. Halhal D, Walters G, Ouazar D, and D.Savic, (1997) Water Network Rehabilitation with a Structured Messy Genetic Algorithm., *J.Water Resources Planning and Management, ACSE*, 123, 137-146.
15. de Schaetzen W, Randall-Smith M.J., Savic D.A., Walters G.A. (1998) A Genetic Algorithm Approach for Rehabilitation in Water Supply. *Proceedings Int. Conference on Rehabilitation Technology for the Water industry*. Lille, France. March 1998.
16. Parker M, Savic D, Walters G, Kappelan Z (2000) SEWERNET: A Genetic Algorithm Application for Optimising Urban Drainage Systems., *International Conference on Urban Drainage via Internet*, May 2000, Prague, CR.
17. Krejci V, Schilling W, Gammer S. (1994) Receiving Water Protection During Wet Weather., *Water Science and Techn.* .29, 219-229.

DEVELOPMENTS IN AGRICULTURE AND THE LOSS OF NATURAL STORMWATER RUNOFF CONTROL IN CENTRAL EUROPE

R.R. VAN DER PLOEG[1], A. CZAJKA-KACZKA[2], M. GIESKA[1], and M. AKKERMANN[1]
[1] Department of Geosciences, University of Hannover, Hannover Germany
[2] Faculty of Earth Sciences, University of Silesia, Sosnowiec Poland

1. Introduction

In the past decade, Germany has been struck by a number of extreme river floods, e.g. of the Rhine in 1993 and 1995, the Oder River in 1997, and the Danube River in 1999. The question as to why the frequency of such events may have increased has been posed repeatedly. The answers that have been given to this question include natural climatic and hydrological fluctuations, climate change, and improper management of waterways and floodplains (see e.g. Immendorf [1]; Marsalek [2]). However, it has also been suggested that changes in land use, accompanied by a partial or total sealing of the land surface, might be responsible for increased surface runoff and peak river discharge. Sieker [3, 4], for example, has pointed at the rapidly increasing urban area in Germany and the high degree of surface sealing in urban areas. He has triggered new developments in urban stormwater control that help to reduce urban peak discharge and flash floods.

In rural areas, surface sealing is less obvious than in urban areas. However, because of its large areal extent, even a partial sealing may be more effective in producing runoff than a total surface sealing in urban areas. In a series of recent articles, van der Ploeg and co-workers have studied postwar changes in land use in Germany (particularly in agriculture) and their possible impact on the environment [5, 6, 7, 8, 9]. These authors conclude that it is likely that postwar changes in land use, particularly in agriculture, contribute to the flood problem that Germany is facing.

Whether or not other Central European countries, like Poland and the Czech Republic, face similar problems, has hardly been studied. Although agricultural issues in these countries have recently received ample attention because of the anticipated east-expansion of the European Union (see, for example, [10] or [11]), environmental aspects in this context usually are not addressed. It is the objective of our present contribution to draw attention to possible environmental problems that may arise when Poland joins the European Union. In particular, we will consider the possible loss of natural stormwater runoff control after Poland starts to intensify its agriculture. To this end, we will review postwar developments in land use and agriculture in Germany,

J. Marsalek et al. (eds.), Advances in Urban Stormwater and Agricultural Runoff Source Controls, 255–266.

256

which (in our opinion) have led to a considerable loss of natural stormwater runoff control and to an increased frequency of extreme river stages and floods.

2. Extreme River Stages in Germany and Poland

In Figure 1, the 11 most extreme stages of the River Rhine at Cologne (9.50 m or more) during the period 1899-1999 are depicted. The figure indicates that in the past century 7 of these were recorded after 1950 and only 4 before 1950. The Danube River at the City of Passau and the Elbe River at the City of Wittenberge feature a similar behaviour ([8]). Hence, it seems that the frequency of extreme stage events on Germany's main rivers has increased during the past half-century.

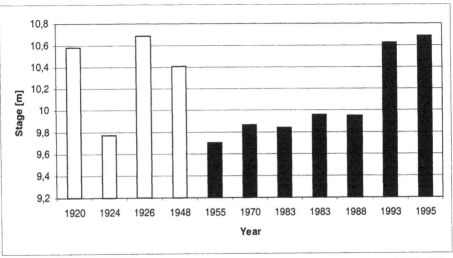

Figure 1. The most extreme stages of the Rhine at Cologne (above 9.50 m) during the period 1899-1999; the white bars denote extreme stages before 1950, the black ones such stages after 1950 [12].

The Vistula River at Warsaw (Poland) shows a different picture. From Figure 2 it can be seen that of the 11 most extreme Vistula stages at Warsaw during the past century (7.25 m or more) only 4 were recorded after 1950, whereas 7 occurred before 1950. Therefore, it seems that no increased frequency of extreme Vistula stages at Warsaw during the second half of the past century can be observed. The discharge behaviour of the Vistula river since 1950, as far as extreme peak flow is concerned, thus differs clearly from that of the Rhine and other German rivers. In our opinion, the increased frequency of extreme river stages and floods in Germany since 1950 can be attributed, at least partly, to changes in land use, particularly in agriculture.

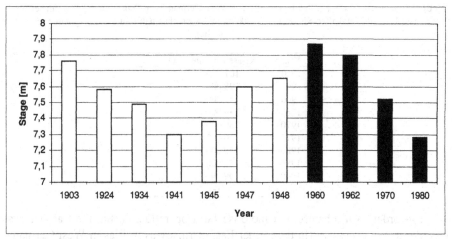

Figure 2. The most extreme Vistula stages at Warsaw (above 7.25 m) during the period 1899-1999; the white bars denote extreme stages before 1950, the black ones such stages after 1950 [13].

3. Changes in Land Use in Germany since 1950

In what follows, changes in land use, particularly in agriculture, in former West Germany since 1950 will be discussed. The data presented are mainly derived from the yearbooks on nutrition, agriculture and forestry, published regularly by the West German Department of Agriculture since 1957 [14]. The early yearbooks, especially the 1957 yearbook, also contain information about land use in the former West Germany before 1957.

Table 1 contains information about a farm consolidation program initiated by the federal government shortly after the Second World War. To save space, only data for each 5th year are shown. The table indicates that the total area affected by the program between 1950 and 1990 was 8,268,000 ha. In view of the total agricultural land area in former West Germany in 1951 (14,371,000 ha), this means that more than 50 % of this area was redivided and reassigned. A main objective of the consolidation program was the enlargement of single farm fields to make them better suited for a mechanised and motorised agricultural management. The consolidation program thus led to a major change in postwar land use in the former West Germany. A far-reaching farm consolidation program was also carried out in the former East Germany. However, not only the average field size increased both in East and West Germany, but also the size of the average farm. In the former West Germany the average farm size was 8.1 ha in 1949, but had increased to 17.7 ha in 1989. In the former East Germany this growth was particularly pronounced [15]. According to Opp [16], the average farm size in Saxony, for example, increased from 7.3 ha in 1955 to 564.7 ha in 1989.

TABLE 1. Selected data (for each 5th year) from the West German Farm Consolidation program between 1950 and 1990; the total affected area (8 268 000 ha) was more than half of the entire agricultural area (14 371 000 ha).

Year	Re-assigned area (ha)
1950	107 000
1955	195 000
1960	241 000
1965	259 000
1970	239 000
1975	216 000
1980	210 000
1985	157 000
1990	99 000
1950-1990	8 268 000 ha

In order to mechanise and motorise farm operations in the field as much as possible, it was necessary (in the former East Germany as well as in West Germany) to drain the vast area of wetland soils. Eggelsmann [17] estimated the total area of wetland soils in former East Germany at 2,400,000 ha or 38 % of the total agricultural land area. For West Germany the same author estimated the area of wetland soils as 4 400 000 ha (or 31 % of the total area in agricultural use), of which two thirds were in need of subsurface drainage. To this end, a government-supported program in the former West Germany was started simultaneously with the farm consolidation program. No records exist on the size of the land area that actually has been drained since 1950. However, the BMELF yearbooks contain detailed information on the public expenditures for civil engineering and drainage work in West German agriculture since 1954. In Table 2 such expenditures are listed. The data in Table 2 can be used to estimate the total land area that was ameliorated in former West Germany since 1950 ([9] [17]). It appears that much, if not all, of the wetland soils in the meantime have been drained, in the former West Germany as well as in East Germany.

TABLE 2. Extent and type of government supported civil engineering and drainage work in agriculture during the period 1954-1990 (after BMELF [14]).

Period	Ditch and Canal Construction	Agricultural Drainage	Subsequent Maintenance Work	Drainage of Uncultivated land	River and Stream Regulations
	------------------------- 10^6 German Mark † -------------------------				
1954 - 1955	125	71	14	10	133
1956 - 1960	452	273	50	35	560
1961 - 1965	802	438	139	33	762
1966 - 1970	911	523	145	20 #	1065
1971 - 1975	897	457	204		1187
1976 - 1980	797	352	277		954
1981 - 1985	554	244	297		960
1986 - 1990	257	95	158		1172
1954 - 1990	4795	2453	1284	98	6793

† All numbers must be multiplied by a factor of 10^6 in order to obtain the costs for each table entry.
# Since 1970 listed together with "Subsequent Maintenance Work"

The large-scale farm consolidation, together with the drainage of the vast area of wetland soils, enabled Germany to mechanise and motorise its agriculture. Indicators of such developments in German agriculture are shown in Table 3. The table shows, for example, that the rate of farm employment in former West Germany dropped between 1950 and 1990 from 5 020 000 to only 833 000. At the same time, the number of tractors increased from only 117 000 in 1950 to 1 276 000 in 1990. The number of other large farm vehicles and machines grew similarly.

TABLE 3. Selected indicators of postwar intensified agriculture in former West Germany [14]

Year	Number of Farm Workers	Horses	Tractors	Combines	Trailers
			10^3 †		
1950	5020 #	1570	117	0.1	
1955	4250 #	1099	404	8	15
1960	3388 (3623 #)	712	797	54	11
1965	2683	360	1106	120	16
1970	2081	253	1335	168	35
1975	1609	341	1401	171	54
1980	1289	382	1417	174	84
1985	1044	369	1426	155	118
1990	833	406	1276	140	139

† All numbers must be multiplied by 10^3 in order to get the proper value for each table entry.
\# Agriculture and forestry combined

However, not only the number of tractors, combines, trailers, and other implements grew, also their size and power increased continuously. As an example, developments with respect to tractors can be considered. In Table 4 some details about the number and power of tractors in postwar West Germany are given. It can be seen that since 1975 the number of tractors hardly changed; their motive power, however, grew steadily.

TABLE 4. Number and power of tractors in postwar West German agriculture

Year	Total number	Number per Power Class	
		<25 h.p.†	>51 h.p.
1949	74 586		
1953	260 548		
1960	797 423	626 038	9137
1970	1 234 968	521 190	82 684
1975	1 287 067	371 000	213 948
1981	1 256 176	223 073	345 948
1987	1 233 201	144 110	477 382
1990	1 156 745	109 187	516 008

† 1 horsepower (h.p.) = 0.7552 kW

With the size and power of the farm implements, also their weight grew continuously. This, in turn, raised in general the axle load of the vehicles. Figure 3 is derived from the work of Sommer [18]. It shows for tractors and combines the general relationship between the motive power and axle load. This development, in general, has led to subsoil compaction.

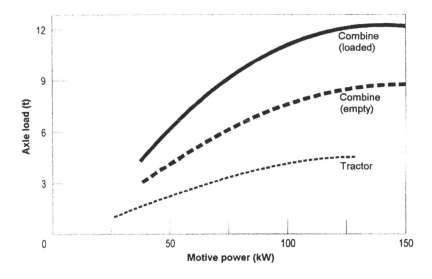

Figure 3. Relationship between motive power and axle load of tractors and combines in Germany [18]

Subsoil compaction is not the only adverse result of intensive agriculture and heavily mechanised soil cultivation. Equally harmful, if not worse, is the physical degradation of the topsoil that affects particularly loess soils in agricultural use. As a consequence of intensive cultivation, soil aggregate stability frequently decreases, aggregates at the soil surface tend to slake during heavy rainfall, and surface sealing and runoff may result.

Postwar developments in agriculture, as just described, are mainly responsible for a reduced potential infiltration of arable soils, especially in the cool and humid winter period. As an example of reduced potential infiltration, experimental work of Schmidt and Stahl [19] can be cited. These authors carried out an infiltration experiment with artificial rain in the state of Saxony. The arable field under study had been cultivated conventionally for many years. Results from Schmidt and Stahl [19], who applied an artificial rainfall rate of 40 mm/h, are depicted in Figure 4. The figure reveals that the conventionally cultivated soil started to show a reduced infiltration only a few minutes after the start of the experiment. The figure also shows that an adjacent field, which had been cultivated in a conservation manner, hardly showed reduced infiltration even after 90 minutes of heavy artificial rain. This field can be considered to represent infiltration conditions that prevailed some 50 years ago, before agriculture was mechanised and motorised. In this respect it is noted that heavy rainstorms, as simulated by Schmidt and Stahl, have repeatedly triggered extreme river stages and floods, especially during winter ([7] [8]).

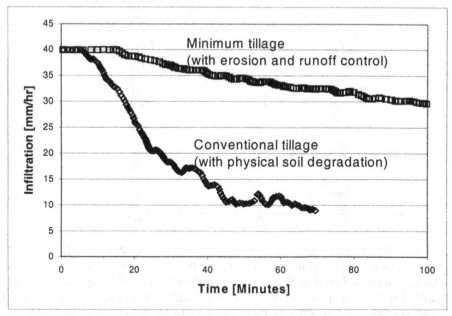

Figure 4. Measured infiltration during heavy artificial rainfall (40 mm/h) on farmers fields, with conventional and minimum tillage, respectively [19].

Physical soil degradation, due to intensive agriculture, may be the main reason for increased surface runoff and peak river discharge, but it is not the only one. In West Germany during the past half-century, an adverse shift in land and soil use has also been observed. The increased urban area, as well as the increased land area used for the growth of small grain, together with a noticeably decreased meadowland area, today produce more runoff during heavy rainstorms in the wintertime, than they did some 50 years ago. Van der Ploeg *et al.* [7] constructed a small hypothetical watershed, that reflects developments in land and soil use in West Germany between 1951 and 1989. Table 5 shows this hypothetical watershed.

In terms of runoff and river discharge, the changes in urban, small grain, and meadowland area are particularly meaningful. Van der Ploeg *et al.* [7] used the US Soil Conservation Service rainfall-runoff model for small agricultural watersheds [20] to evaluate such changes. They concluded that these postwar changes in land and soil use may have a substantial impact on the occurrence of extreme stages of rivers such as the Rhine.

TABLE 5. Land use in former West Germany in 1951 and 1989, visualised by means of a 1000ha hypothetical watershed.

Land use	Area (ha) 1951	Area (ha) 1989
Urban area	74	122
Fallow	2	7
Row crops	89	91
Small grain	185	223
Forage	45	12
Pasture	80	91
Meadow	157	108
Farmsteads, country roads	20	5
Forest	283	298
Heath, bogs	10	6
Water	18	18
Wasteland	37	19
Total	1000	1000

Besides physical soil degradation and changes in land and soil use, the extensive postwar drainage of agricultural land possibly has also contributed to the problem of increased river peak discharge and floods. Van der Ploeg and Sieker [9] showed that the area in former West Germany with a subsurface drainage system possibly comprises 3,000,000 ha or more. Because the tile drainage system, in general, has been designed with narrow spacing, the release of subsurface water in times of excessive rain or snowmelt is large. The vast, highly drained agricultural land area thus may contribute substantially to river peak discharge and floods in times with excessive rain and soil wetness. For details, we refer to van der Ploeg and Sieker [9].

Summarising, we thus can state that due to postwar developments in land use, particularly in agriculture (physical soil degradation, shifts in grown crops, and drainage), Germany seems to have lost much of its natural stormwater runoff control. Although climate changes and adverse hydraulic constructions along its main rivers may play a role, we hypothesise that these postwar developments in agriculture also contribute substantially to the problem of increased extreme river stages and floods. It is likely that the increased frequency of extreme river stages in Germany, exemplarily depicted for the River Rhine in Figure 1, is related to these developments.

4. Land Use and Agriculture in Poland

Statistical data on Poland's agriculture have recently become easily accessible [21]. Land use conditions in Poland can now be compared with those of postwar Germany. In Table 6 such a comparison is made. It can be seen that the land use in Poland in 1997 resembled that of West Germany in 1951; the proportion of total urban area in Poland is a little smaller and that for agricultural areas a little larger.

TABLE 6. A comparison of land use in Poland in 1997 and in former West Germany in 1951 and 1989.

Land use	Poland 1997	West Germany 1951†	West Germany 1989
Urban (inc. road system)	6.4	7.4	12.2
Agriculture	60.0	57.8	53.7
Forest + Woodland	28.0	28.3	29.8
Uncultivated Land	1.0	1.0	0.6
Water	3.0	1.8	1.8
Wasteland	1.6	3.7	1.9
Sum	100.0	100.0	100.0

† With estimates for West Berlin and Saarland

In Figure 5 the growth of the average farm size in former West Germany after the Second World War is shown. Also, the present farm size in Poland is given. As shown in Table 6 with respect to land use, Poland's present farm size resembles that of the former West Germany in the early 1950s.

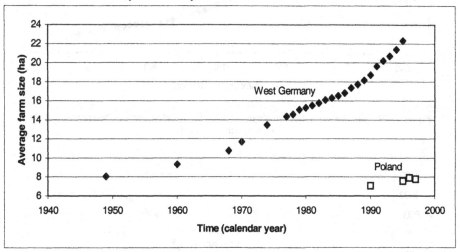

Figure 5. The average farm size in former West Germany after the Second World War and the present farm size in Poland.

The rate of employment in agriculture shows a similar picture (Figure 6): The present rate of employment in agriculture in Poland (27 %) is even higher than that in West Germany around 1950 (25 %).

To a lesser degree the same holds for the motive power in agriculture. Figure 7 shows developments in postwar West Germany and the present degree of tractor power in Poland. It appears that present-day land use conditions in Poland resemble those of West Germany in the 1950s and 1960s. This, incidentally, is also true for other agricultural parameters, such as fertiliser use and crop yields. From Figure 2, Table 6, and Figures 5-7, as well as from other data that we do not show, it appears that postwar developments in agriculture, as observed in Germany, so far have not taken place in Poland. Consequently, the agricultural land area in Poland still retains considerable natural stormwater runoff control. Accordingly, the Vistula River at

Warsaw does not show the postwar increased frequency of extreme stages, that characterises the discharge of the Rhine at Cologne. However, it must be appreciated that things will change, after Poland joins the European Union and as soon as the country starts intensifying its agriculture. Unless Poland's agriculture is modernised wisely, the country will experience the same agriculturally-induced environmental problems that Germany is facing today. The same is possibly also true for the Czech Republic.

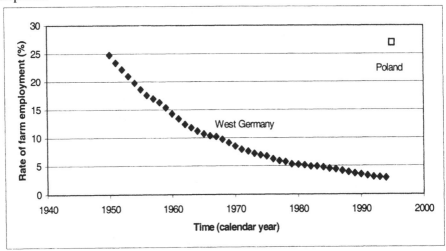

Figure 6. The rate of postwar farm employment in the former West Germany and present-day farm employment in Poland.

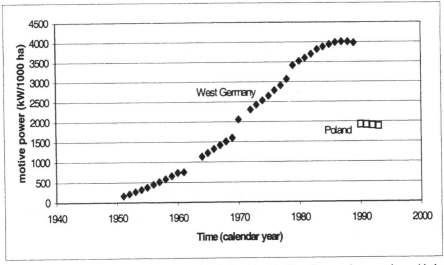

Figure 7. Postwar motive force per 1000 ha of agricultural land in West Germany in comparison with the corresponding parameter of mechanisation for Poland in 1997.

5. Conclusions

It is likely that postwar developments in land use and agriculture in Germany have led to the increased formation of surface runoff and peak river discharge in wet periods. It is also likely that such developments in Poland so far have not taken place. However, if Poland starts intensifying its agriculture in the same way this was done in Germany, it may experience the same environmental problems that Germany is facing. This may be true also for other Central European countries.

6. Acknowledgement

This research was supported by the Federal Foundation for the Environment (Deutsche Bundesstiftung Umwelt) at Osnabrück.

7. References

1. Immendorf, R. ed. (1997) *Hochwasser: Natur im Überfluß?* C.F. Müller Verlag, Heidelberg.
2. Marsalek, J. et al., eds. (2000) *Flood issues in contemporary water management*. NATO Science Series (2. Environmental Security, Vol. 71), Kluwer Academic Publishers, Dordrecht.
3. Sieker, F. (1996a) Dezentrale Regenwasserbewirtschaftung als Beitrag der Siedlungswasserwirtschaft zur Hochwasserdämpfung. Zeitschrift Stadtentwässerung und Gewässerschutz **34**.
4. Sieker, F. (1996b) Dezentrale Regenwasserbewirtschaftung in Siedlungsgebieten: Ein Beitrag zur Dämpfung extremer Hochwasserereignisse ? Geowissenschaften **14(12)**, 531-538.
5. van der Ploeg, R.R., Ringe, H., Machulla, G., and Hermsmeyer, D. (1997) Postwar nitrogen use efficiency in West German agriculture and groundwater quality, *J. Environ. Qual.* **26**, 1203-1212.
6. van der Ploeg, R.R., Ehlers W., and Sieker F. (1999) Floods and other possible adverse environmental effects of meadowland area decline in former West Germany, *Naturwissenschaften* **86**, 313-319.
7. van der Ploeg, R.R., Hermsmeyer, D., and Bachmann, J. (2000) Postwar changes in land use in former West Germany and the increased number of inland floods, in J. Marsalek et al. (ed.) *Flood issues in contemporary water management*, NATO Science Series (2. Environmental Security, Vol. 71), Kluwer Academic Publishers, Dordrecht, pp. 115-123.
8. van der Ploeg, R.R., Ilsemann J., Hermsmeyer D., and Machulla G. (2000) Eine geänderte Landnutzung in der Nachkriegszeit als Mitverursacher der Hochwasserprobleme in Deutschland? in S. Heiden, R. Erb und F. Sieker (Hrsg.) *Hochwasserschutz heute - Nachhaltiges Wassermanagement*. Erich Schmidt Verlag, Berlin. (in press).
9. van der Ploeg, R.R. and Sieker F. (2000) Bodenwasserrückhalt zum Hochwasserschutz durch Extensivierung der Dränung landwirtschaftlich genutzter Flächen, *Wasserwirtschaft* **90(1)**, 28-33.
10. Hausner, J. and Marody, M., eds. (1999) *Three Polands: The potential for and barriers to integration with the European Union*, Friedrich Ebert Foundation, Warsaw.
11. Quaisser, W., Hartmann, M., Hönekopp, E., and Brandmeier, M. (2000) *Die Osterweiterung der Europäischen Union: Konsequenzen für Wohlstand und Beschäftigung in Europa*, Friedrich-Ebert-Stiftung, Bonn.
12. Vogt, R. (1995) Hochwasser in Köln: Erfahrungen, Maßnahmen, Schlußfolgerungen, Ratgeber "*Mit dem Hochwasser leben*" **1**, 48-55. Stein-Verlag, Baden-Baden.
13. IMiGW (Instytut Meteorologii i Gospodarki Wodnej). (1948, ..., 1983) *Hydrological Yearbook of Vistula Catchment*. Transport and Communications Press, Warsaw (in Polish).
14. BMELF (Bundesministerium für Ernährung, Landwirtschaft und Forsten). (1957,... 1991) *Statistisches Jahrbuch über Ernährung, Landwirtschaft und Forsten*. Verlag Paul Parey, Hamburg und Berlin (1957,... 1975), Landwirtschaftsverlag GmbH, Münster-Hiltrup (1976,... 1991).

15. Ministerrat der DDR (1966, ..., 1989) *Statistisches Jahrbuch der Land-, Forst- und Nahrungsgüter-wirtschaft.* Staatliche Zentralverwaltung für Statistik, Berlin (Ost).
16. Opp, Chr. (1998) Geographische Beiträge zur Analyse von Bodendegradationen und ihrer Diagnose in der Landschaft. *Leipziger Geowissenschaften,* Band 8.
17. Eggelsmann, R. (1971) *Umfang und bodenbedingter Bedarf der Landeskultur-Maßnahmen in der Bundesrepublik Deutschland.* Z. f. Kulturtechnik u. Flurbereinigung **12**, 153-162.
18. Sommer, C. (1994) Belastung, Beanspruchung und Verdichtung von Böden durch landwirtschaftliche Maschinen, *Landbauforschung Völkenrode,* Sonderheft 147.
19. Schmidt, W. and Stahl H. (1999) Erosionsstrategien in Sachsen als Beitrag zur Wertschöpfung in der Landwirtschaft. S. 176-182. In: Dekan der Landwirtschaftlichen Fakultät (Hrsg.) *Wertschöpfung in der landwirtschaftlichen Primärproduktion* (Wissenschaftliche Beiträge der 7. Hochschultagung der Martin-Luther-Universität Halle-Wittenberg, 16. und 17. März 1997 in Leipzig).
20. Schwab, G.O., Fangmeier D.D., Elliot W.J., and Frevert R.K. (1993) *Soil and Water Conservation Engineering,* 4th ed. John Wiley and Sons, Inc., New York.
21. GUS (Glowny Urzad Statystyczny) (1991, ..., 1999) *Statistical Yearbook of Poland,* GUS, Warsaw (in Polish).

INFLUENCES OF LAND USE AND LAND COVER CONDITIONS ON FLOOD GENERATION: A SIMULATION STUDY

D. NIEHOFF[1] & A. BRONSTERT[2]
[1]*Potsdam Institute for Climate Impact Research*
P.O. Box 60 12 03, D-14412 Potsdam, Germany
[2]*University of Potsdam, Institute for Geo-Ecology*
P.O. Box 60 15 53, D-14415 Potsdam, Germany

1. Reasons for the Study

Both the landscape and the river systems in large parts of Central Europe have undergone major changes in the past, and there is no doubt that these environmental changes have altered the nature of floods in this region. But due to the complexity of the processes involved, the *magnitude* of their impact on *storm-runoff generation* and subsequent *flood discharge* in the river system is still uncertain. This uncertainty offers a vivid platform for various contradictory opinions on this topic, quite often disregarding *scale-dependencies* and *boundary conditions* in a way that ensures public attention rather than relevance.

The work presented in the following focuses on three main questions, strictly referring to the spatial scale and the boundary conditions for which statements are made:

(1)   Which runoff generation mechanisms are likely to be affected by land-use and land-cover changes at the mesoscale?
(2)   To what degree can flooding be *mitigated* by water retention measures in the landscape at the mesoscale?
(3)   How does the influence of land-use and land-cover changes on storm-runoff generation depend on *catchment characteristics* and *spatial scale* as well as *event characteristics* and *temporal scale*?

The investigation does not address the influences of river training conditions and retention along the river courses on flood-wave propagation.

267

J. Marsalek et al. (eds.), Advances in Urban Stormwater and Agricultural Runoff Source Controls, 267–278.
© 2001 *Kluwer Academic Publishers. Printed in the Netherlands.*

2. Study Set-Up

This study investigates the impact of land-surface conditions on storm-runoff generation at the lower mesoscale. For this purpose, three catchments (100 to 500 km²) within the Rhine basin have been selected which represent different characteristic land-use patterns with either dominantly urban, agricultural or forest structure. The hydrological response of these catchments to heavy rainfall is simulated for present land use as well as for various possible future land-use conditions, involving both process-oriented hydrological modelling and the delineation of land-use scenarios.

At the time when the project started, there was no satisfactory method available for the generation of spatially distributed land-use scenarios. Therefore this task is performed using a new approach, the land-use change modelling kit LUCK [1]. LUCK operates on the basis of gridded maps and provides a method for the spatial transformation of overall land-use change trends into spatially distributed land-use patterns. The overall trends (scenario targets) are obtained from external analysis as percentile amounts of change, in most cases lumped for administrative units. Land-use conversion is realised on the basis of an evaluation of the local site characteristics of each grid cell as well as its neighbourhood relationships.

In order to quantify the influences of land use on storm-runoff generation, the deterministic and distributed hydrological model, WaSiM-ETH, has been chosen, because of its well-balanced mixture of physically based and conceptual modelling approaches. The model, which was developed by Schulla [2], originally focused on simulating the influence of climate change on the catchment water balance. This expresses itself in a sophisticated handling of meteorological data, spatial discretisation and land-use parameterisation. In order to improve the representation of land-use characteristics with respect to runoff generation, the soil module has been extended, now explicitly allowing for macropore infiltration, siltation and crusting of the soil surface, connection of sealed surfaces to the sewer system and decentralised retention in the landscape. The underlying mechanisms of these extensions are outlined in *Section 3*. A more detailed description can be found in [3].

3. Model Extensions

3.1 MACROPORE FLOW

Within the extended WaSiM-ETH, macropores are treated as a single linear storage which interacts with the soil matrix. *Infiltration* into the macropores occurs as *infiltration excess* or as *saturation excess* of the soil matrix. *Exfiltration* out of the macropores into the soil matrix is limited to unsaturated conditions and is controlled by the *saturation deficit* of the matrix and the *storage coefficient* of the single linear storage. This procedure is a simplification of an approach which was developed by Bronstert [4] for the microscale. Macropore storage capacity is given as the product of macroporosity and the average depth of the macropore layer. As information on the interaction between macropores and the soil matrix is sparse, the macropore storage coefficient is expected to be subject to calibration.

The influence of macropores on storm-runoff generation depends enormously on *rainfall characteristics* and antecedent *soil moisture* conditions. This is illustrated by *Figure 1*, which compares a flood resulting from a *convective* storm event to one induced by an *advective* storm event. The convective event is characterised by high rainfall intensities and low antecedent soil moisture, whereas the advective event with much lower intensities is accompanied by high antecedent soil moisture. The simulations refer to the Lein catchment with an area of 115 km² in southwest Germany. Its hydrological behaviour is characterised by a gently undulating terrain and a loess cover with a thickness of up to 20 m. At some places, the underlying strata of marl and gypsum reach the surface. The two flood events are of the same order of magnitude, both having a return period of about three years.

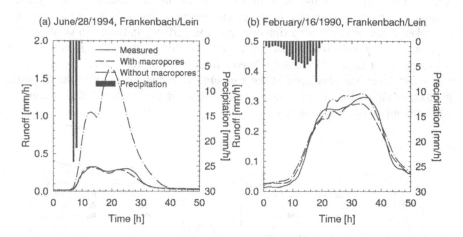

Figure 1 Simulation of two flood events in the Lein catchment (115 km²) with a return period of approximately three years as a response to (a) a convective and (b) an advective storm event, showing the impact of macropores on runoff generation

As *Figure 1 a* demonstrates, generation of infiltration-excess overland flow and the resultant direct runoff is by far overestimated when infiltration is calculated solely as micropore infiltration controlled by soil hydraulic conductivity. This is also true for the advective event (*Figure 1 b*), but because of lower precipitation intensities and higher soil moisture the consequences are less obvious. In this context it is important to notice that for advective events like the one in February 1990, which contributed to a large flood in the Rhine basin, infiltration-excess overland flow is of very minor importance anyway.

3.2 SILTATION EFFECTS

Due to their characteristic particle size distribution, loess soils are generally susceptible to aggregate breakdown and siltation during high intensity rainfall, resulting in surface sealing and a drastic *decrease in hydraulic conductivity* at the soil surface as well as a *decline in macropore connectivity* (e.g. Römkens *et al.* [5]). In the past, the impact of siltation on infiltration and the production of infiltration-excess overland flow has been studied extensively with the help of sprinkler experiments at the plot scale. The reduction of hydraulic conductivity obtained from such experiments is in the range of one order of magnitude (e.g. Roth *et al.*, [6]). Much less is known about the magnitude of macropore disconnection due to siltation and the impact of siltation on runoff generation at the catchment scale.

The siltation module, which has been developed as an extension for WaSiM-ETH, takes into account the decrease in hydraulic conductivity at the soil surface as well as the reduced inflow of infiltration excess into the macropores. Hydraulic conductivity of the soil surface is reduced by simply multiplying it with a factor C_{silt} that takes values between zero (impermeable soil surface) and one (no siltation). C_{silt} depends on the *precipitation intensity*, the actual *canopy cover* and a given maximum reduction factor C_{max}, which can be derived from experimental studies. Once a poorly permeable layer has developed at the soil surface, *regeneration* from the aggregate breakdown takes place over a longer period of time (up to several months) or abruptly, when sowing or harvesting is done.

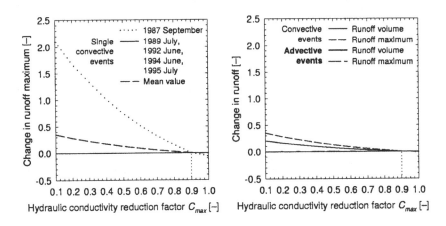

Figure 2 Sensitivity analysis for the hydraulic conductivity reduction factor Cmax used by the siltation module within the extended WaSiM-ETH

When applying the siltation module to the Lein catchment (see *Section 3.1*), with parameter values as they were measured at the plot scale, an unrealistically high amount of infiltration-excess is simulated and catchment runoff is drastically overrated. *Figure 2* contains the results obtained from a *sensitivity analysis* conducted with data from the Lein catchment for the factor C_{max}. The sensitivity analysis was done separately for five *convective* and six *advective* storm events which induced floods with return periods between two and eight years. The right-hand side of *Figure 2* shows mean values for the impact of C_{max} on the runoff volume and runoff maximum, and suggests a moderate increase of catchment runoff for convective storm events with high precipitation intensities and no significant impact for advective storms. The diagram on the left, on the other hand, demonstrates that only one convective event met the siltation criteria of *high precipitation intensities* associated with a *low canopy cover* that occurred, because cereals had been harvested shortly before. For this event, sensitivity of C_{max} is so extraordinarily high, that even for a moderate two-fold maximum reduction of hydraulic conductivity ($C_{max} = 0.5$) the peak runoff is nearly doubled. In contrast to the exaggerating simulations, only 6% of the precipitation contributed to the flood event as it was recorded in September 1987. This value is just as low as for the other convective events. These findings support the observation of Roth *et al.* [6] that a large percentage of the infiltration-excess generated locally *re-infiltrates* in areas which are not affected by siltation. Consequently, the empirical evidence obtained for siltation at the plot scale cannot simply be adopted in order to describe runoff generation at the catchment scale.

3.3 SEALED SURFACES AND URBAN SEWER SYSTEMS

Urban areas, on the one hand, consist of asphalt or paved surfaces which allow only very little infiltration and are often connected to a sewer system. On the other hand, they contain greens, parks, green strips or gardens, where higher infiltration and soil storage conditions can be found. Lumping soil parameters in these areas inevitably leads to an overestimation of the influence of built-up areas on storm runoff generation and underrates the compensating effect of green areas within settlements.

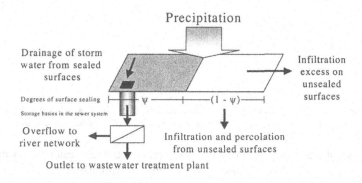

Figure 3 Model concept for sealed surfaces within a grid cell and their connection to the sewer system

In order to take into account this pronounced form of *heterogeneity* within grid cells (often referred to as *subgrid variability*), each grid cell is divided into a sealed and an unsealed part according to the *degree of sealing* typical for the cell actual land-

272

sewer system. Another advantage of this procedure is that it allows for a distinction between densely settled areas and low density settlements. But this approach is not limited to settlement areas; linear infrastructure like roads within grid cells with agricultural or forest land use can also be considered in this way. Usually, such linear landscape elements are disregarded, because they most often do not appear in gridded maps with cell sizes of 100 x 100 m and more.

4. Hydrological Simulation of Land-Use Scenarios

4.1 URBANISATION SCENARIOS

As an example of the impact of urbanisation on storm runoff generation, in the following, a simulated response to an increase in settlement and industrial area of 10% and 50%, respectively, is described. In the Lein catchment (see *Section 3.1*), such an increase corresponds to a growth of these land-use types from 7.4% of the catchment area to 8.1% and 11.1%, respectively.

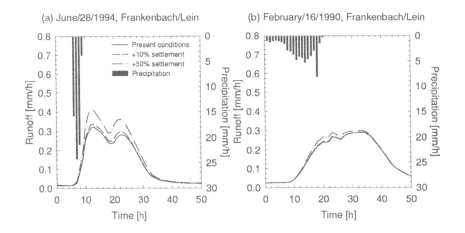

Figure 4 Simulation of two flood events in the Lein catchment (115 km²) as a response to (a) a convective storm event and (b) an advective storm event for present conditions and two urbanisation scenarios

Figure 4 is a comparison of two flood events in the Lein catchment which have already been described in *Section 3.1*. It contains simulation results for present conditions as well as for the two urbanisation scenarios. The comparison demonstrates that the increase in flood volume and peak runoff due to urbanisation is much more distinct for the *convective* storm event than for the *advective* one, although the precipitation volume as well as the peak flow are of the same order of magnitude for both events and represent a return period of approximately 2 to 3 years in both cases. The markedly slighter effect on the advective event is the result of (1` ' ' ' *antecedent soil moisture* which balances differences in soil characteristics as well as (2) *lower precipitation intensities* which prevent an overflow of the sewer system.

Table 1 Increase in runoff volume and peak flow due to a 50% growth of the settlement and industrial areas in the Lein catchment; the events are sorted by the urbanisation impact on runoff volume

Year, month	Increase in runoff compared to present conditions		Simulated baseflow contribution to volume [%]	Duration [h]	Return period approx. [y]
	Maximum [%]	Volume [%]			
1990, February	3.4	3.7	19	150	2
1993, December	5.9	2.7	17	250	8
1997, February	3.9	2.7	19	150	7
1982, December	1.7	1.5	27	225	3
1983, May	0.6	0.9	39	300	4
1988, March	0.0	0.0	52	650	3
Mean	2.6	1.8	29	290	4.5

This argumentation is also supported by a comparison of various advective events with different return periods, as shown in *Table 1*. The comparison reveals a strong correlation between the impact of urbanisation on runoff and the *baseflow contribution* to the flood event, which serves as an indicator for high groundwater levels and high soil moisture.

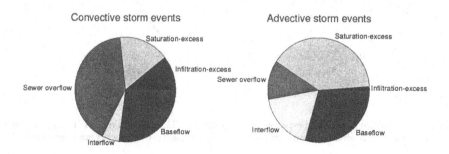

Figure 5 Runoff components simulated with WaSiM-ETH for the Lein catchment for five convective and six advective storm events with return periods between two and eight years

Another comprehensive source of information are the runoff components that are simulated for the different flood events (*Figure 5*). The two pie charts reveal pronounced differences in the *dominating runoff-generation mechanisms* depending on the event characteristics (rainfall intensities and antecedent moisture conditions). The response to convective storm events is dominated by *sewer overflow* from sealed surfaces as well as a considerable amount of *infiltration-excess* mainly from

274

agricultural areas. In contrast to that, for advective events *subsurface flow* processes and *saturation excess* prevail.

4.2 URBAN STORMWATER MANAGEMENT SCENARIOS

Following the *Action plan on flood defence* of the International Commission for the protection of the Rhine [7], between 1998 and 2020, the sealing of 1.3% of the catchment area of the Rhine river caused by asphalt or other paved surfaces is to be compensated by decentralised infiltration measures in settlement areas. Transferred to the Lein catchment, the rainwater from 54% of the area that is impervious and connected to a sewer system would have to be collected in *infiltration ponds* instead of becoming combined sewage. These 54% refer to 6.9% of settlement areas within the catchment with a degree of imperviousness of about 0.35.

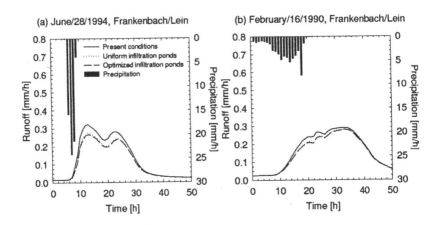

Figure 6 Simulation of two flood events in the Lein catchment (115 km²) as a response to (a) a convective storm event and (b) an advective storm event for present conditions and two storm-water management scenarios

Figure 6 illustrates the effect of two scenarios: The first one includes infiltration measures in all the settlement areas. The second one excludes the 25% of the settlement areas with the highest groundwater levels from the connection to infiltration measures; at the other sites the efforts are intensified to reach the same target as in the first case.

Due to the same causal relationships which were discussed for the urbanisation scenarios, the effect of *infiltration measures* in settlement areas is stronger for *convective* storm events with high infiltration intensities and lower antecedent soil moisture than it is for *advective* events with low precipitation intensities and under wet antecedent conditions (see *Figure 7*). Compared to the scenario which provides all settlement areas with infiltration measures, the simulation results for the second scenario show only a slight improvement. The concentration of deeper infiltration ponds on a smaller area leads to a faster saturation of the soils below these ponds, which counterbalances the advantage of the sites with principally lower groundwater levels.

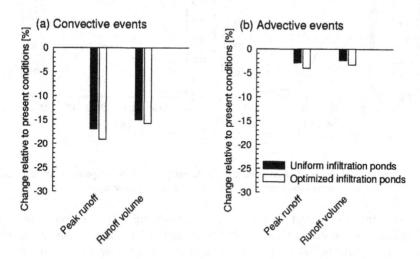

Figure 7 Simulated change in peak runoff and runoff volume in the Lein catchment due to infiltration measures in settlement areas, given as mean values for (a) five convective and (b) six advective storm events with return periods of two to eight years

4.3 AGRICULTURAL MANAGEMENT SCENARIO

According to the European Union's agricultural strategy paper *Agenda 2000*, a 10% reduction of the agricultural production area is strived for in the near future. This target has directly been transferred to the Lein catchment by changing the sites least suitable for agricultural production to *temporary set-aside areas*.

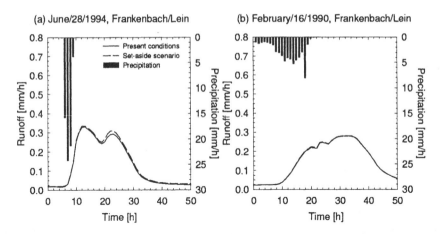

Figure 8 Simulation of two flood events in the Lein catchment (115 km²) as a response to (a) a convective storm event and (b) an advective storm event for present conditions and a set-aside scenario

The simulated impact of this scenario on storm runoff generation is marginal for both events illustrated in *Figure 8*. The slight increase in runoff visible in *Figure 8 a* is due to a probably less dense canopy cover on the set-aside fields at the end of June. Nearly no change in runoff is calculated for the *advective* event shown in *Figure 8 b*. This is also realistic, because soil storage conditions are not substantially affected by this measure, whereas possible modifications of the infiltration conditions are not relevant during rain storms with low precipitation intensities. However, the uncertainty inherent in the land use parameterisation of the hydrological model is far larger than the simulated change, which is being speculated on.

5. Conclusions

Modelling the influence of land use and land use changes on storm runoff generation is highly dependent on an adequate consideration of the following factors and aspects:

(1)　*Land-cover* characteristics and their influence on the appearance of the soil surface.
(2)　*Runoff generation processes* including the influence of the soil surface on infiltration.
(3)　*Spatial distribution* of land-use types and land-cover characteristics.
(4)　*Temporal and spatial dynamics* of storm events.
(5)　*Initial and boundary conditions*, particularly regarding antecedent moisture conditions.
(6)　*Spatial and temporal scales*, for which the model is designed and/or applied to.

WaSiM-ETH has been chosen for this study because it considers the spatial distribution of catchment characteristics and spatial and temporal dynamics of climate variables in a sophisticated manner. The model extensions that are presented here were developed in order to improve the representation of the land cover and the unsaturated zone within the model. Thereby, the additional parameters and process descriptions have led to an increase in model uncertainty. On the other hand, the influences of land cover characteristics on storm runoff generation are now represented in a comprehensive way that allows to track complex interactions and coherences, which otherwise would not be obvious. Despite the uncertainty associated with the simulation results, some general conclusions can be drawn from what has been learnt during the model development and application:

(1)   The influence of land use and decentralised flood mitigation measures on storm runoff generation is stronger for convective storm events with high precipitation intensities than for long advective storm events with low precipitation intensities.

(2)   Yet convective storm events are of very minor relevance for the formation of floods in the large river basins of Central Europe, because usually they are restricted to local occurrence.

(3)   Precipitation volume as well as antecedent soil moisture conditions and groundwater levels are of major importance for the degree, up to which land use and decentralised flood mitigation measures can influence storm runoff generation. The magnitude of a flood peak or the return period of a flood event, respectively, are less meaningful indicators in this respect.

(4)   Disastrous flood events in the large river basins in Central Europe often are the result of a coincidence of flood events in a great number of subcatchments. But the floods do not necessarily have to be disastrous in the subcatchments themselves. Therefore the conclusion that the influence of land use is principally low for large floods in large basins is not valid.

6. Acknowledgements

This work was funded by the European Union as a contribution to the INTERREG Rhine-Maas Activities (IRMA), and by the German Federal Environmental Agency (UBA).

7. References

1. Fritsch, U., Katzenmaier, D. & Menzel, L. (1999) Land use scenarios for flood risk assessment studies, in Musy, A., Pereira, L.-S. & Fritsch, M. (eds.) *Emerging technologies for sustainable land use and water* (2. Inter-Regional Conference on Environment-Water, Sept. 1–3, 1999 at EPFL, Lausanne, Switzerland), Lausanne, Switzerland.

2. Schulla, J. (1997) *Hydrologische Modellierung von Flussgebieten zur Abschätzung der Folgen von Klimaänderungen* (Zürcher Geographische Schriften, Heft 69), Zurich, Switzerland.

3. Katzenmaier, D., Fritsch, U. & Bronstert, A. (2000) Quantifizierung des Einflusses von Landnutzung und dezentraler Versickerung auf die Hochwasserentstehung. Chapter of a book to be published by the German Environment Foundation (DBU), *in press*.

4. Bronstert, A. (1999) Capabilities and limitations of detailed hillslope hydrological modelling, *Hydrological Processes* **13**, 21–48.

5. Römkens, M.J.M., Luk, S.H., Poesen, J.W.A. & Mermut, A.R. (1995) Rain infiltration into loess soils from different geographic regions, *Catena* **25**, 21–32.

6. Roth, C.H., Helming, K. & Fohrer, N. (1995) Surface sealing and runoff generation on soils derived from loess and pleistocene sediments, *Zeitschrift für Pflanzenernährung und Bodenkunde* **158**, 43–53.

7. IPCR (International Commision for the Protection of the Rhine) (1998) *Action plan on flood defence*, Koblenz, Germany.

SOURCE CONTROL MEASURES FOR STORMWATER RUNOFF IN URBAN AREAS

Investigation of Impacts on Floods in the Saar Catchment

F. SIEKER and U. ZIMMERMAN
University of Hannover, Institute of Hydrology, Water Resources and Agricultural Water Management
Appel Str. 9a, 30167 Hannover, Germany

1. Introduction

Stormwater runoff in urban areas has been handled up to now mostly according to the principle: "Stormwater runoff has to be discharged as fast and as completely as possible in order to avoid or to diminish damages or inconveniences caused by flooding or water ponding within settlements". This principle is implemented in the design of sewer systems, either combined or separated. Although sewer systems successfully achieved the protection aims in the past, their disadvantages increased recently. The overloading of sewer systems as a consequence of connection of developed areas leads to a cost intensive extension of sewer systems and wastewater treatment plants. Further, there is a loss of groundwater recharge in urban areas. And last but not the least, the impacts of sewer systems on the receiving waters regarding quantities and pollution loads became more important and cannot be tolerated any longer. Consequently, such impacts have to be reduced. Hence, the attitude towards stormwater runoff is going to change. Application of source control measures instead of mere stormwater discharge wherever possible is going to become the new principle.

First, objectives and features of the new operation principle are described. Second, a new system to handle road runoff by source control measures is described to prove that the principle can be applied in practice. Third, some preliminary results of a research project concerning effects of flood reduction by source control measures within the catchment of the River Saar are presented.

2. Objectives and Features of the New Principle

First, it has to be emphasised that the present level of protection of urban areas from flooding, water ponding and other inconvenience has to be maintained in the case of application of source control measures. But an additional aim of the principle of on-site source control measures is to:

> "Match the water balance equation of the urbanised state of the catchments to the pre-urbanised state as best as possible"

J. Marsalek et al. (eds.), *Advances in Urban Stormwater and Agricultural Runoff Source Controls*, 279–285.
© 2001 *Kluwer Academic Publishers. Printed in the Netherlands.*

This approach is mainly directed to the surface runoff component and groundwater recharge. To approach the evaporation component of the natural state after urbanisation is difficult.

For non-urbanised areas, most processes of the water balance take place within natural soils. Consequently, for an urbanised area, the soils of the unpaved parts should be used preferably for the management of stormwater runoff. The unpaved part of urban areas available for this purpose is mostly relatively small and inconveniently positioned with respect to the runoff producing areas. Arranging the installation of on-site source control measures with these constraints is the required task.

To achieve the aim of managing stormwater runoff, an on-site stormwater management system has been developed. This system can be adapted to local conditions like soil conditions (especially hydraulic conductivity), the available size of unpaved area, different levels of safety standards against overloading and other factors. This system, called "Swale-Trench-System" or "Trough-Trench-System", is described in detail by Sieker [1, 2]. The main functions of the system are as follows:

- treatment of polluted runoff by surface infiltration and percolation through the topsoil layer,
- storage of the first flush of runoff within the grass swale (trough) on the surface,
- underground storage in the pore volume of a trench, filled with coarse gravel, lava or synthetic material etc.,
- groundwater recharge by infiltration from the trench into the surrounding soil, and
- discharge of throttled runoff by small drainage pipes.

In general, the system consists of different swale-trench elements connected by drain pipes located on private sites or along public roads. Depending on the hydraulic conductivity of the given soil, the system can be simplified to single swale-trench elements without connecting pipes or to swales without any trenches below, for instance in the case of sandy soils. The system can be tailored in different technical ways suited to the local circumstances and to the imagination of the designer.

3. The Road Draining System INNODRAIN

In order to prove that the principle of the swale-trench system is a generally applicable alternative to conventional sewer systems, the following technical version of the system, called INNODRAIN, is described. It has been developed especially for narrow roads within residential areas and to calm traffic. Figure 1 shows the details. In order to reduce the infiltration area within the road, the upper part of the system, the trough, is reduced to 4-5 % of the connected road area. The inside surface elevation of the trough is about 20 cm below the road surface level. The trough is constructed of concrete frame elements of different form, size and number. It is fed by road runoff from the gutter via a sedimentation inlet in order to prevent entry of the coarse material. The topsoil layer of the trough is about 25 cm deep and is densely planted with grass or shrubs. In order to avoid road flooding, an overflow pipe is installed to conduct overflow water directly into the drainage pipe system in the underground. The trench is filled with coarse gravel or other material with high pore volume.

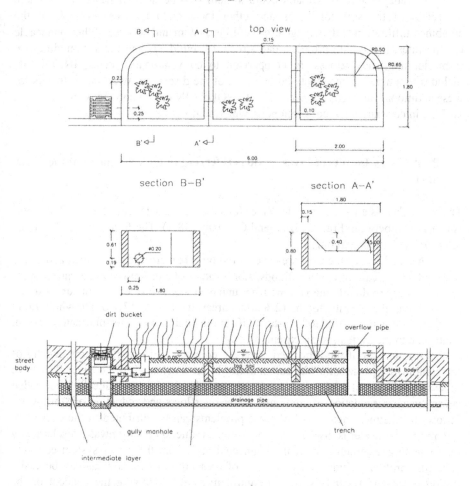

Figure 1. Innovative road draining system INNODRAIN.

The following parameters can be changed and adapted to adjust the water balance to that of natural areas:

- size and depth of the infiltration area (the inner area of the trough),
- permeability of the topsoil layer,
- overflow frequency of the overflow pipe,
- cross-section and length of the trench,
- type of trench material regarding different specific pore volume, and
- throttle rate of the drainage pipe.

By adapting these variables, the system can be applied nearly everywhere regardless of the soil conditions and other local circumstances. Because of the combined infiltration and drainage effects of the system and because of the changeable throttle rates, the system protects road structures, buildings and the surrounding soil from flooding and wetness. In comparison to conventional systems, INNORAIN includes functions of treatment and retention of the discharge. Where investments for these additional functions are necessary, INNODRAIN can be applied with savings of 20-30% in comparison to conventional sewer systems.

4. Preliminary Results of Investigations referring to the Catchment of the River Saar

The River Saar is a tributary of the River Moselle. The catchment of about 7,600 km^2 is distributed between France (52%) and Germany (48%). The following explanations refer to the German part only.

About 12 % of the total area is covered by urban areas; 5% is paved and 7% is unpaved. In the case of extreme floods, stormwater volume from paved areas amounts approximately to double the volume from unpaved areas (inside and outside of urban areas) so that the contribution of 12% urban areas amounts to 17% of the whole flood volume. So it is interesting to investigate if source control measures in urban areas can contribute to the mitigation of extreme floods.

About 90% of the urbanised areas are connected to sewer systems, mostly of the combined type. Most of the communities in the catchment of the River Saar have to solve problems at present regarding the rehabilitation of hydraulically overloaded sewer systems, reduction of combined sewer overflows and reduction of stormwater inflows to treatment plants. All of these problems can be attributed to the cause that too much stormwater is discharged. Therefore, a reduction of stormwater discharge by disconnecting a certain part of the connected area from the sewer system can help solve the problems. Indeed, a good deal of presently connected areas can be easily disconnected and treated by source control measures following the concept of the swale-trench system. The decision about the realisation of the measures depends on their costs compared with the costs of conventional solutions of the problems. Conventional methods to solve the problems mentioned above are: replacement of overloaded sewers, construction of retention tanks and basins, and investments in treatment plants.

As examples, three communities of about 5-10 thousand inhabitants located in the catchment of the River Saar are investigated regarding that comparison. Steps of the investigations are as follows.

- Compute the demand for renovations of the existing sewer system by long-term simulation of the discharge in the sewer system in order to find out locations and frequencies of flooding for the given state.
- Conduct long-term simulations of the pollution in combined sewer overflows for the given state, in order to estimate the volumes of retention tanks required by the existing technical guidelines.
- Estimate the costs of replacement of overloaded reaches of the sewer systems and estimate the costs of retention tanks required in the case of application of the technical guidelines.
- Conduct a GIS-based investigation of the urban areas to develop detailed maps of the possible source control measures divided in steps of possible realisation.
- GIS-based description of the types of source control measures referring to the different steps of the swale-trench system. Make a distinction between source control measures with and without discharge throttling.
- Repeat the first two steps referring to acceptable disconnection scenarios – long-term simulation of remaining flooding within the sewer system and long-term simulation of remaining overflow events regarding different possible disconnection scenarios.
- Compare the latter results with the targets of technical rules. In cases where the targets are not achieved, additional conventional measures must be taken into account.
- Estimate the costs of the source control measures and, if necessary, the costs of the additional conventional measures required.
- Compare the costs and benefits of non-monetary aspects of, for instance, groundwater recharge and flood reduction. For similar costs, the latter aspects should be the decisive factors.

Results of the investigated examples show first that the potential of easily convertible source control measures of the swale-trench system is generally sufficient to compensate the effects of the conventional methods and to achieve the aims of the technical rules. Second, the results show that the costs of the source control concept are lower than the costs of the conventional way of renovation the draining system. The portion of easily convertible paved area connected to existing sewer systems can be estimated on average at 25%.

Considering now the effects of source control measures on groundwater recharge and reduction of floods in rivers, it can be noticed that these effects are free of additional costs. But the question arises if the source control measures of the swale-trench system, which is normally sized by using input data for intense but short urban rainfall events, are also able to process rainfall periods of days or even weeks which are mostly responsible for extreme floods in rivers.

In order to investigate this question, typical elements of the swale-trench system, in combination with typically occurring permeability coefficients of a given soil, were exposed to a rainfall period of about 30 days and a precipitation depth of about 300 mm in total. A similar rainfall caused a large flood in the River Saar in 1995. Table 1 shows the results.

TABLE 1. Retention effects of different source control measures on extreme floods.

On-site measure	Infiltration capacity (m/s)	Stored water volume	
		Portion of rainfall (%)	In m^3 for a 50 ha area; 294 mm rainfall
Swale (Trough)	1×10^{-5}	100	147,000
Swale (Trough)	5×10^{-6}	95	139,000
Swale (Trough)	1×10^{-6}	56	82,000
Swale-trench elements non-connected	1×10^{-6}	93	122,000
Swale-trench system, Elements connected	5×10^{-7}	72	106,000
Swale-trench system, Elements connected	1×10^{-7}	21	30,000

In the catchment of the River Saar, the hydraulic conductivity of the given soil is estimated to be 5×10^{-7} to 1×10^{-6} m/s (clay, loam) on average. The data in Table 1 indicate that 72 – 93 % of the stormwater runoff from paved areas can be retained by source control measures of the swale-trench system. So it can be seen that source control measures, which are primarily installed cope with urban drainage problems, are also able to handle long-term rainfall periods which cause extreme floods. This provides an additional reason for application of source control measures instead of only conventional sewer systems.

Significant effects of urban source control measures on river floods can be expected only if they are applied to all the urban areas in the catchment. Referring to the catchment of the River Saar (the German part), the effects of reduction may be estimated at 3% of the flood volume for the case where all the potential urban areas have been converted (long-term implementation). This value may not seem to be very much, but for the German part of the Saar catchment of about 3,650 km^2 and with a rainfall depth of about 300 mm (the 1995 value), a reduction of 3% of runoff is equivalent to a retention volume of about 33 million m^3. It is very difficult (and expensive) to find a place to construct a retention basin of 33 million m^3 within a heavy urbanised area like the German part of the Saar catchment. And it must be kept in mind that source control measures are financed by urban investments and therefore free of costs relating to flood reduction effects.

5. Conclusions

Source control measures for stormwater runoff can be summarised by the concept of the swale-trench system. The system can be applied independently of soil conditions and other local circumstances. It can be applied in different technical ways depending on the local conditions, the objectives, and the imagination of the designer. For the management of road runoff a special technical version of the system, called INNODRAIN, has been developed. Investigations in several communities have shown that the swale-trench system and its variations can be successfully applied in existing urban areas in order to solve problems resulting from hydraulically overloaded sewers, reduce combined sewer overflows and for other benefits. Comparisons with conventional solutions have shown that applications of the system reduce costs. Last but not the least, extended applications of source control measures in urban areas can contribute to flood mitigation.

6. References

1. Sieker, F. (1998) On-site stormwater management as an alternative to conventional sewer systems: a new concept spreading in Germany, *Water, Science and Technology* **38 (10)**, 65-71.
2. Sieker, F. (2000) Investigations of the effects of on-site stormwater management measures in urban and agricultural areas on floods, in J. Marsalek (ed.) *Flood Issues in Contemporary Water Management*, Kluwer Academic Publishers, Dordrecht, pp. 303-310.

CONSERVATION TILLAGE – A NEW STRATEGY IN FLOOD CONTROL

W. SCHMIDT, B. ZIMMERLING, O. NITZSCHE, & S.T. KRÜCK
Sächsische Landesanstalt für Landwirtschaft
P.O. Box 221161
04131 Leipzig
Germany

1. Introduction

It may seem at first that there is no connection between tillage methods on arable land and flood disasters in a watershed. But flood disasters are frequently the consequence of extensive amounts of water originating from surface runoff from soils due to a lack of infiltration caused by soil sealing or crusting. The last has to be seen in context with soil erosion on arable land, which results from inhibited water infiltration through soil siltation. Soil sealing is caused by raindrops hitting the soil surface with a force great enough to destroy soil aggregates. Dispersed surface clods and aggregates form a thin sealing soil layer, which inhibits water infiltration in a very efficient way [1]. On sloped arable land, inhibited infiltration by soil sealing causes surface water runoff, which causes on- and off-site damages through soil erosion.

The best way to decrease or to prevent surface runoff on arable land is to prevent soil sealing and crusting. In the following we want to show that conservation tillage combined with mulch seeding is one of the most efficient strategies against siltation on arable land. Conservation tillage has an influence on a number of physical and hydrological soil parameters. In most cases, this contributes to a drastic reduction of surface runoff on arable land.

In Saxony more than 60 % of arable land (450,000 ha) is endangered and regularly afflicted by water erosion. To reduce or prevent on- and off-site damages caused by water erosion, extensive soil protection measures are needed in entire regions or watersheds. Conservation tillage and mulch seeding are recommended as effective methods against water erosion by the agricultural extension service and will be more and more practised on arable land in the future.

Since reduced water erosion is closely connected with reduced water runoff, conservation tillage on arable land in the whole catchment may be both an effective strategy against water erosion and an efficient element of flood control. In the following this possible relationship will be demonstrated by field experiment results. These results were obtained at various sites in Saxony in different tillage systems with simulated rainfall experiments (Fig. 1).

J. Marsalek et al. (eds.), Advances in Urban Stormwater and Agricultural Runoff Source Controls, 287–293.

Rainfall simulation setup			
- Plot size:	1 m^2	1 m^2	44 m^2
- Rainfall intensity:	0.7 mm*min^{-1}	1.9 mm*min^{-1}	0.7 mm*min^{-1}
- Duration:	60 min	20 min	60 min
- Nozzle type:	VeeJet80/100 (1 unit)	VeeJet80/100 (1 unit)	VeeJet80/100 (15 units)

Examined parameters

- Surface water runoff, infiltration
- Soil loss
- Soil aggregate stability [2]
- Soil organic matter content
- Soil coverage with mulch

Figure 1.   Experimental methods applied in experiments assessing the effect of heavy rainfall events on physical and hydrological soil parameters

2. Soil Tillage Systems in Germany

Cultivation of annual crops (e.g., wheat, barley, oil-seed rape, sugar beet, corn) in Germany is presently achieved with three different tillage systems:

1. Soil tillage with the mouldboard plough, defined as *conventional tillage*. It is characterised by a soil-turning action to a depth of up to 30 cm. This is highly effective in burying and thereby killing annual and perennial weeds and volunteer crops. Ploughing produces a clean surface, which facilitates precision seeding with common seeding machines.
2. Soil tillage without mouldboard plough, defined as *conservation tillage*. This includes shallow tillage methods without the soil turning action of the plough. Different tillage implements are used, like cultivators, (rotary) harrows, disks, normally in conjunction with mulch seeding of different crops (for example, corn, sugar beet, oil-seed rape, wheat, barley).
3. No-tillage systems or *direct seeding*. Except for nutrient injection and seeding, soil is usually not disturbed between harvesting and planting under no-till systems.

The main tillage method currently practised in Germany is ploughing. This is done to kill annual and perennial weeds as well as volunteer crops and/or to incorporate organic residues and manure into the soil. Another reason is the preparation of an even and clean surface, which facilitates precision seeding [3].

For ecological (see below) and economical reasons, there is an increase in application of conservation tillage methods. Direct seeding is practised with some crops (e.g., wheat after oil-seed rape), but because of unresolved problems, such as straw management and weed control and seeding, it is still rarely practised.

3. Conservation Tillage and Water Infiltration

Results presented in Table 1 and Figures 2 and 3 demonstrate that surface runoff is increased after conventional tillage. This is due to the bare soil surface after the protective cover of mulch or crop residues have been incorporated into the soil by the plough. Consequently, raindrops exert direct destructive impacts on soil aggregates on the surface during heavy rain storms. This induces soil sealing and crusting. In addition, permanent ploughing directly and indirectly promotes excessive pan formation and subsoil compaction, and interrupts macropores produced by earthworms or plant roots. All this inhibits water infiltration [3]. Reduced or prevented water infiltration then leads to water runoff and water erosion on sloped land.

TABLE 1. Water infiltration, runoff, and soil loss after conventional and conservation tillage during two subsequent (dry and wet) heavy rainfall events with a duration of one hour. (Plot size: 44 m², rainfall intensity: 0.7 mm*min⁻¹, duration: 60 min)

		Rain simulation (dry)			Rain simulation (wet)		
Texture	Tillage	Infiltration [% of rainfall]	Runoff	Soil loss [t*ha⁻¹]	Infiltration [% of rainfall]	Runoff	Soil loss [t*ha⁻¹]
Silty	Convent.	55.4	44.6	82.8	32.9	67.1	45.3
loam	Conserv.¹	92.1	7.9	0.2	70.9	29.1	1.2
Loam	Convent.	94.0	6.0	0.1	63.8	36.2	0.7
	Conserv.²	99.9	0.1	0	97.4	2.6	0.02

Mulch coverage: 1) 60 %, 2) 10 %

TABLE 2. Mulch cover, soil organic matter, aggregate stability, infiltration rate and soil loss by water runoff after conventional tillage, conservation tillage (8 years) and direct seeding (8 years). (Plot size: 1 m², rainfall intensity: 0.7 mm*min⁻¹, duration: 60 min)

		Conventional tillage	Conservation tillage	Direct seeding
Mulch cover	[%]	1.0	30.0	70.0
Soil organic matter	[%]	2.0	2.6	2.5
Aggregate stability	[%]	30.1	43.1	48.7
Infiltration rate	[%]	49.4	70.9	92.4
Soil loss	[g*m⁻²]	317.6	137.5	33.7

On the other hand, conservation tillage combined with mulch seeding ensures high infiltration rates and thereby reduces direct water runoff and consequently soil erosion during a sequence of heavy rainfall events on different soils. This positive effect is due to the mulch layer (Table 2), which restrains the force of raindrops hitting the soil surface, thereby protecting the soil aggregates. Also, conservation tillage increases the content of soil organic matter and biological activities in the top 5 to 10 cm of soil. This has a positive effect on the soil aggregate stability (Table 2). Increased earthworm activity provides for further stability. Due to the ingestion of mineral and organic matter by the earthworms [4], organomineral complexes are formed and deposited on the soil surface.

Macropores or biopores are crucial for rapid drainage of rainwater into the soil during heavy rainfall. These are constructed by anectic earthworms or left behind by dead roots. For infiltration the continuity of these pores from the soil surface into the subsoil is essential [5]. Ploughing regularly destroys the macropores down to the ploughing depth, so that at any given time only newly constructed pores are available. On the other hand the soil stratification and the macropores are conserved when conservation tillage methods are used.

In addition, the mulch layer of organic crop residues, left on the soil surface when conservation tillage methods are applied (Table 2), serves as food source for anectic earthworms [4]. This food supply and the reduced soil disturbance promote the development of earthworm populations under long-term conservation tillage, thus allowing for a further increase in macropores.

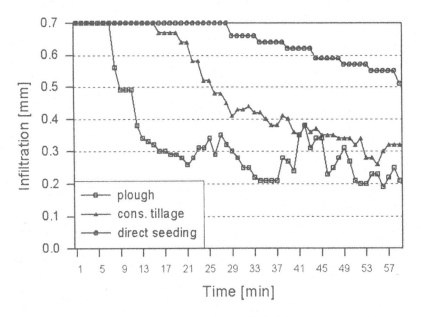

Figure 2.   Course of infiltration on arable land (texture silty loam) after conventional tillage and conservation tillage (8 years). (Plot size: 1 m², rainfall intensity: 0.7 mm*min⁻¹, duration: 60 min)

Altogether, the effects of conservation tillage lead to an increase in infiltration (Tables 1 and 2). The results presented in Figure 2 clearly show the differences in infiltration during the course of the irrigation on crop land cultivated with different tillage methods. The rapid increase in water runoff on the ploughed site, as opposed to the continuous high infiltration rates on sites with conservation tillage and direct seeding, is conspicuous (Table 2, Fig. 2). As shown in Figure 3, sites with conservation tillage have a high infiltration capacity during a longer period of time, even under the circumstances of unusually heavy rainfall events (e. g. 38 mm over a period of 20 min) as compared to ploughed sites.

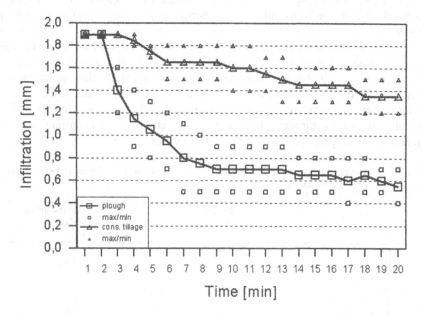

Figure 3. Course of infiltration on arable land (texture silty loam) after conventional tillage and conservation tillage (2 years). (Plot size: 1 m², rainfall intensity: 1.9 mm*min⁻¹, duration: 20 min)

4. Conservation Tillage and Flood Control

Observations of the siltation of the soil surface and the associated effects observed in irrigation experiments with heavy rainfall exhibit good repeatability. Siltation reduces water infiltration into the soil and causes water runoff even before the topsoil is saturated so that the water holding capacity of the soil is not utilised. Rapid surface runoff can contribute to river flooding.

On the other hand, higher infiltration rates on arable land cultivated with conservation tillage methods can sustainably reduce water runoff by improving the utilisation of the water holding capacity of the soil (Tables 1 and 2, Figs. 2 and 3). Although for a small area rapid drainage of the infiltrated water through compacted deeper soil layers or drainage systems can not be excluded, on the larger scale of a catchment area it can be assumed that a distinct time lag is achieved in drainage which can be important for the course of a flood.

The infiltration rates and their variation in time, shown in Tables 1 and 2 and in Figures 2 and 3, were assessed in irrigation experiments with a rainfall simulator on differently cultivated sites with a maximum irrigated area of 44 m². On the basis of data collected in these plot experiments and applying a water balance simulation model, these concepts are presently being evaluated in a research project, which is financially supported by the German Federal Foundation for the Environment. The main issue is to establish to what extent widespread conservation tillage contributes to the suppression of flooding. In this context, a broad implementation of conservation tillage in the whole catchment area is an important requirement to influence flooding in this way.

It can be assumed that this could be achieved in Saxony. Due to the fact, that vast areas are endangered by water erosion (60 % of the arable land, equivalent to 450.000 ha) it is required that conservation tillage and mulch seeding, which promote infiltration and erosion control, be implemented on a broad scale. Mulch seeding techniques are, therefore, promoted by the agricultural extension services and are financially supported through the program for environmental farming methods (Programm Umweltgerechte Landwirtschaft) of the Saxony state.

TABLE 3.   Development of the mulch seeded area due to the program for environmental farming methods in Saxony from 1993/94 till 1998/99

Year	Arable land [ha]	Proportion of total arable land in Saxony [%]
1993/1994	4.146	0.6
1994/1995	27.096	3.8
1995/1996	44.585	6.1
1996/1997	54.188	7.5
1997/1998	74.721	10.3
1998/1999	79.084	10.9

After five years of operation of the environmental farming program, in 1998/99 the mulch seeded area amounted to nearly 80,000 ha and covered 10.9 % of the arable land in Saxony (Table 3). Presently conservation tillage methods are applied to various crops on about 30 % of the arable land in Saxony, in individual years. Durable conservation tillage with mulch seeding in the course of a complete crop rotation is presently practised on single farms, with each farm covering a maximum of 3000 ha.

These figures show, that conservation tillage and mulch seeding are already applied in Saxony to an extent at which an influential effect on flooding in catchment areas can be expected.

5. Summary and Future Prospects

Since arable land usually comprises the main portion of a river catchment area, flooding and water protective cultivation methods are of great importance. Converting from conventional plough tillage to conservation tillage with mulch seeding drastically changes a number of physical and hydrological soil parameters. Especially, the vulnerability to soil erosion, surface water runoff and nutrient leaching from the surface layer are reduced by these cultivation systems. As confirmed by a number of irrigation experiments and soil loss measurements in Saxony, the use of conservation tillage on a large scale has the potential to reduce flooding and to reduce the contamination of surface waters. Therefore, conservation tillage with mulch seeding is recommended to farmers and is financially supported.

Due to the effect on erosion control, conservation tillage methods are mentioned in the German soil protection legislation as important measures for soil conservation on farmland. Since the resulting application of conservation tillage methods is also profitable for farmers, it can be assumed that the use of this cultivation method will expand. To quantify the effects of extensive conservation tillage on the course of flooding events, further development of water balance simulation programs is needed.

6. References

1. West, L.T., Chiang, S.C. and Norton, L.D. (1992) The Morphology of Surface Crusts, in M.E. Sumner and B.A. Stewart (eds.), *Soil Crusting - Chemical and Physical Processes*, Lewis Publishers, Boca Raton Fla., pp. 73-92.
2. Murer, E.J., Baumgarten, A., Eder, E., Gerzabek, M.H., Kandeler, E. and Rampazzo, N. (1993) An improved sieving machine for estimation of soil aggregate stability (SAS), *Geoderma* **56**, 539-547.
3. Ehlers, W. and Claupein, W. (1994) Approaches toward conservation tillage in Germany, in M.R. Carter (eds.), *Conservation Tillage in Temperate Agroecosystems*, Lewis Publishers, pp. 141-165.
4. Topp, W. (1981) *Biologie der Bodenorganismen*, Quelle und Meyer, Heidelberg.
5. Beisecker, R. (1994) Einfluss langjährig unterschiedlicher Bodenbearbeitungssysteme auf das Bodengefüge, die Wasserinfiltration und die Stoffverlagerung eines Löß- und eines Sandbodens, in H.-R. Bork, H.-G. Frede, M. Renger, F. Alaily, C. Roth and G. Wessolek (eds.), *Bodenökologie und Bodengenese*, TU Berlin, Berlin, Heft 12., pp. 1-195.

STORMWATER SOURCE CONTROL AND PUBLIC ACCEPTANCE

An application of adaptive water management

G.D. GELDOF
Tauw, Water and Spatial Planning, PO Box 133, 7400 AC Deventer &
Faculty of Technology & Management, University of Twente, PO Box
127, 7500 AE Enschede
The Netherlands

1. Introduction

In the late 1980s, source control was practically unheard of in the Netherlands. Urban water management was perceived as synonymous with sewer system management. Hardly any attention was paid to surface waters in the city, and just about the only residents who made use of it were grandmothers taking their grandchildren to feed the ducks. Measures taken with regard to sewer systems were in a maintenance context or consisted of end-of-pipe control. However, all this changed in the early 1990s. Source control was put on the agenda by, among others, the Socoma working group of the Joint Committee of Urban Storm Drainage.

The first projects involving source control on a larger scale (most of which were located in newly built urban areas) were launched in 1993. At the outset, subsurface infiltration facilities were the only instrument used; other methods followed later. At the time, the key issues were of a technical nature: would the infiltration of rainwater contaminate the soil? Would it result in clogging? Could infiltration be applied in areas with a high groundwater level? What materials should be used? Would they remain operational at low temperatures in wintertime? Would infiltration contribute to the abatement of depletion? Would it help reduce peak discharges? Would source-control techniques, from an emission-abatement point of view, be a valid alternative to storage settling tanks? Would infiltration mean savings on sewage treatment plants? In the meantime, research findings and – most of all – practical experience have yielded the answers to all these questions, and to many more.

Source control has been fully accepted since 1995. The results obtained from field studies have removed the last obstacles to its application. Practically all newly built residential areas are now equipped with a sustainable urban water system. The Fourth National Policy Paper on Water Management (1998) – which formulates the official Dutch water management policy – makes it mandatory to apply source-control measures in newly built urban areas. It is no longer acceptable to simply pump the stormwater run-off from paved surfaces into sewage treatment plants together with wastewater.

J. Marsalek et al. (eds.), Advances in Urban Stormwater and Agricultural Runoff Source Controls, 295–303.
© 2001 *Kluwer Academic Publishers. Printed in the Netherlands.*

In the years to come, the real challenge will be posed by existing urban areas, where considerably less progress has been made in introducing source-control measures. The obstacles are no longer exclusively of a technical nature: legal, financial and communicative matters are now also an issue, as are social matters. Nowadays, the use of technology needs to be fine-tuned to the needs of society. The work done by different disciplines must be integrated in order to achieve success, and such integration is not easy to realise.

2. Different Worlds

Existing urban areas have various disciplines working in them. On the one hand, technicians are in charge of constructing new infrastructure and managing and maintaining public spaces. If sewer pipes need to be replaced, roads are broken up and new pipes are laid. If drivers take short cuts and drive too fast through residential areas, speed ramps are built. On the other hand, there is also plenty of work for social scientists. They co-ordinate district redevelopment and initiate activities aimed at fostering social coherence and a sense of security. They also counsel neighbourhood interest groups.

In the past, the two disciplines operated independently of each other. They did not need each other and perceived their respective disciplines as being diametrically opposite. And in fact, they are paradigmatically different from each other. Thus, both groups found themselves dug in in their respective trenches. While technicians mocked social scientists for their alleged vagueness and irrelevance, the latter accused the former of being inflexible and insensitive to change. As a result, the no man's land between the two disciplines was not a place anyone wanted to be.

Figure 1. The 'trenches' dug by social and technical scientists, with no man's land between them.

The approach to urban areas is changing rapidly. Where parties used to act and react from their own point of view, they are now expected to anticipate matters and to co-operate very closely. Although an integrated approach benefits the quality of the living environment, source-control techniques cannot be integrated by technicians working in isolation: such measures are diffuse and small scale by definition, and cannot be implemented without the co-operation of the local residents. It is imperative to bridge the gap between technicians and social scientists in order to implement source-control measures in existing urban areas.

3. Adaptive Management

It will take time for the two parties to climb out of their trenches, although they have begun to emerge. There is now growing interest in a concept known as adaptive management [1], which involves not two but three parties, with the third one being positioned somewhere in between. What used to be no man's land has been discovered by a few pioneers to be a world of complexity (see e.g. [2]). It is a fascinating and attractive world in which instead of making a compromise between technology and social processes, they are merged. Complexity manifests itself between too much and too little, between order and chaos, between certainty and uncertainty. Necessity and coincidence take turns. The essence of adaptive management is that efforts are aimed at making manageable the complexity within that grey area, rather than counteracting it.

The world of complexity is a meeting place for people from various disciplines. It is a breeding ground for synthesis, one which offers promising opportunities to manage the application of source control.

Figure 2. The discovery of complexity.

This paper outlines ways in which to deal with the involvement of citizens. It illustrates how complexity can be made manageable. To start with, we look at what is common practice among today's technicians. We then highlight a few characteristics of adaptive management, and provide an example of how such can be put into practice.

4. Need for Involvement

A key characteristic of source control is that, the prevention is better than cure principle. Thus, the number of large-scale facilities will decrease while the emphasis is shifted to small-scale decentralised facilities. As a consequence, the interests of water managers will extend to the smallest parts of the water system. In order for small-scale measures to be successful, large numbers of people and groups of people must co-operate. They need to be persuaded to rally behind the formulated policy and to demonstrate their commitment – and it must never be taken for granted that people will actually do so. The mere publication of a convincing memorandum explaining the various water management plans will not persuade people to saw off their downpipe in order to have the rainwater from their roof run into their garden. It takes a lot more than that.

5. Creating Public Support

Water managers believe that it is possible to create the public support required for water management measures. Their argumentation is that the government defines objectives to be met by water management, and these objectives are considered as an indisputable fact. In order to meet these objectives, the bodies in charge of water management are obliged to conceive sets of measures that represent the best possible solution. Obtaining maximum results at minimum cost is what the water manager owes the taxpayer – the citizen. If the set of measures does yield maximum results at minimum cost, then it may objectively be regarded as the optimum. However, the implementation of the measures will not always be the exclusive responsibility of the water management body: the co-operation of other parties, including citizens, is an absolute prerequisite, which means that public support must be created. Good information will create awareness among the public; it will stimulate active support and make people ready to take action.

Many of the elements of this line of argumentation are understandable. Good information, for example, is of the utmost essence. But the arguments put forward are also a little naive: they are based on the assumption that individual citizens feel some sort of commitment to water management because such is in the interest of the collective – and this is far from the truth. Hardly anyone perceives water management as something worthy of their commitment. To most people, it simply is not an issue – until their basement is flooded, their gutters leak or the water from the tap comes out brown. Nowadays, the average citizen is swamped with appeals for all kinds of good causes: runaway children, criminality, social segregation, chronic illnesses, traffic safety, alcoholism, education, famines and the ever-increasing number of overstressed people, etc. These are very important matters, so what would make water management such a special issue that people should find it in themselves to approach it with extra commitment?

This is why the key line of argumentation that has governed daily practice so far must be called into question with some urgency, particularly in circles that tend to underestimate the challenge of creating public support.

6. Why People Tend not to Spontaneously Jump with Joy

We know from our experience in urban redevelopment and the management of public open spaces that to most people water is not an issue. Anyone making an effort to involve himself or herself in urban redevelopment processes will find that the really hot items are the lack of parking spaces, drivers taking short cuts through residential areas, the shortage of playgrounds for children and the abundance of dog faeces. These items stir up emotions. Water may be interesting, but it touches a tender spot in only a very few people. Consequently, few visit their local water board building on its open day, and even fewer participate in the elections for a new board member. For many years, the public has been kept at a distance from anything related to water management. The basic message sent out by the government was 'Don't worry, we'll take care of you'. Professional water management experts ensured that the water from our taps was fresh and that the level of surface waters always remained under control. We are now experiencing the backlash. It will take a while for the public to shift from years of non-involvement to commitment. Involvement does not develop overnight, nor is it delivered on demand.

7. Accelerating the Process does not Work

It is clearly understood that policy objectives will not be achieved if people are not committed. Especially the abatement of depletion and non-point sources, for which broad public support is a prerequisite, is a difficult process. At present, there is a tendency not to discuss the current daily practice, but to strengthen it and speed it up. This is counterproductive. By artificially maintaining the old practice, the principle of more haste, less speed is engendered. Processes become tense and success does not materialise.

8. A Different Approach

As stated, adaptive management offers good prospects because it does not communicate in only one direction, as the previously described practice does. Operations must be carried out interactively, and such interaction must not be limited to planning that involves the main stakeholders, but should primarily be aimed at concrete projects in the immediate living environment of people. Important focal points are as follows [3].

1. Dare to be inconsistent.
2. Leap before you look.
3. Support those who deal with concrete matters.
4. Do not avoid conflicts.

These four points are briefly explained below.

Re 1.   When searching for solutions, uniformity and standardisation are no longer the aim. It is diversity and identity that are appreciated in the urban area; and because each situation is unique, the risk of creating a precedent is small. Especially where sustainable development is the goal, the wheel may be reinvented again and again.

Re 2.   All the matters stated above are closely connected and, for the most part, the processes involved in urban water management are unpredictable. The idea of actually finding optimum solutions in advance in order to present them to the involved public – so as to obtain its support – does not work. In such a complex work situation, learning by doing is the preferred mode of operation, in which information, experience and public support run parallel to each other. People concerned should be given a chance to live through a design process instead of having to accept a design that was drawn up prior to the creation of public support and then executed.

Re 3.   Differences in views that have a high level of abstraction often seem to be irreconcilable. A well-known example is the unremitting discussion between municipalities and water boards about the assignment of tasks in the management and maintenance of watercourses. The municipality sees to the drainage, and the water board sees to its discharge. For over 15 years now, the point of discussion has been where drainage stops and discharge begins. In practice, however, agreement could easily be attained in 90% of cases, as each specific situation exacts its own logic. Those working in the field are able to convert abstract ideas into concrete 'products' and should be more appreciated. Too many people are sitting at a computer devising optimum solutions from too great a distance.

Re 4.   Often, a consensus is striven for in which conflicts are seen as ripples in the smooth surface of the optimum solution. Change is not possible this way. It is better to reveal and even emphasise points of conflict as early as possible, so that all arguments can be thoroughly debated.

Water managers will remain responsible for the direction in this approach to adaptive management. Changes can be made quickly, as long as they are given enough time.

9. Trust and Mutual Dependency

Time after time, trust has proven to be of importance. If involved parties do not have confidence in each other, public support evaporates. Trust is an important building block of the social capital.

Processes in residential areas often prove that the public's confidence in the government is low. Meetings with the community sometimes end in slanging matches. If confidence is low, resistance will increase, often fed by simplistic or even incorrect information. On top of that, people do not forget negative experiences. For example, in Biezenstraat in Nijmegen, the municipality proposed infiltrating rainwater from paved surfaces into the subsoil. This evoked fierce reactions during a community meeting. Some residents were afraid that their basement would become flooded. The discussion showed that the residents had little faith in the municipality. A brothel at the end of the street was causing quite some nuisance in the area, mainly due to the extra cars that were parked there. The municipality had not dealt with complaints adequately in the past and the problem continued to exist, according to the residents. Resisting the infiltration of rainwater was a way for them to express their dissatisfaction.

It is therefore essential to win the confidence of the community and thus lay the foundation for constructive co-operation. With an adaptive approach, the involved agents act on the basis of equality. Hierarchical structures are less important than network relations, and the initiators of the process (in urban water management, mostly the municipality, water board and province) look for points of mutual dependency. It is clear that the collective authorities depend on the residents for a source-related approach for water. The residents, on the other hand, will have many requests with regard to parking, traffic short cuts, playgrounds for children and other aspects of liveability within the public urban space. In turn, they largely depend on the collective authorities. Therefore, the parties are mutually dependent. By viewing water as an essential part of the living environment and not as a separate issue, it becomes possible to create win-win situations.

10. De Vliert

The adaptive process is illustrated by de Vliert in Den Bosch, a residential area largely built in the 1930s and home to about 5,000 inhabitants. De Vliert exemplifies how objectives in urban water management can be realised in existing urban areas. The municipality was the initiator, together with De Maaskant water board. The main reason to start a project in de Vliert was the need to replace the sewer system. Usually, this entails digging up the streets, replacing the sewers and then resurfacing. Such a project means a great deal of nuisance for the residents without a visible improvement as recompense. In this case, however, the municipality believed that the residents were entitled to some visible improvements if work of this nature was to be carried out. In addition, any opportunity to improve the water system should be made use of.

The municipality and the water board started by drawing up a concept of desirable improvements of the water system, which befits their authority and competence. This concept was then embedded in a municipal sewer plan. The authorities opted for a source control-related approach, which involved making as much use as possible of the rainwater in and round the house, or else infiltrating it into the soil. The use of durable materials and an increased awareness of the residents can cause a reduction of the number of non-point sources in the surface water. There was a clear understanding that such a concept is realistic only if the process is characterised by openness and reciprocity. It is simply not possible to force residents to cut their downpipes in two in order to infiltrate rainwater into their own property unless they benefit from this directly. Another point clear to the authorities was that the confidence the residents had in them was low. On many occasions in the past, the municipality had not exerted itself to deal with requests and comments from the residents, and realised that this sudden change in policy could arouse suspicion. Confidence had to be restored.

This process was started in de Vliert in September 1997 with an information evening in the community centre De Slinger. Despite the fact that Ajax was playing that evening, the turnout was good: over a hundred residents attended, the slight majority of whom were women. That evening, the municipality (a representative of the water board was present but did not play an active role) presented its concept and put forward some ideas. However, it deliberately did not present a plan; in fact, there was no plan. The local residents were asked to make designs. They could also sign up for a walk through their area, which was to take place a week and a half later on a Saturday afternoon, and be followed by a workshop where they would be able to make the designs. There was a lot of scepticism at the beginning of the evening. Such comments as 'The municipality already has a plan anyway!' and 'Where can I get a complaint form if things go wrong?' were made. Nevertheless, a rather sudden turn for the better occurred about thirty minutes into the discussion. It seemed as though the negative reactions of the 'notorious complainers' had triggered an extra positive attitude from other local residents. Many of them sensed that the officials present really meant what they were saying this time, and thus were annoyed by the residents who tried to ridicule their ideas. In the end, about 25 residents signed up for the walk and workshop, and this was the maximum number indicated by the municipality.

Such a notion as integrated water management does not really interest people unless it is put in the context of their own living environment. During the workshop, designs were made for four areas by means of the principle calculation, design and argument. The designs may not have been completely up to the mark with regard to water management, but they matched the demands of the residents regarding parking, combating traffic taking short cuts through the area, the quality of local green areas, playgrounds for children and even senior citizen's schemes. People were actually thinking together. Ideas were also put forward for a work of art featuring a rainwater column. The column would slowly release its water to a little mill. If it were raining, the mill would turn quickly. The longer it did not rain, the slower the mill would turn. After a dry spell lasting a couple of weeks, it would stop completely. Such a piece of art does not have any environmental output but is very valuable for residents.

In the design phase, the residents provided some basic ideas for a design 'atlas'. In a later phase, these ideas were distributed in the area in consultation with experts. The solutions and techniques often varied from street to street. In one street, for instance, the residents were against the infiltration ditch (*wadi*) planned for the grass strip between street and pavement, because they feared it would lead to inconvenience when getting in and out of a parked car. Infiltration ditches were not acceptable there.

The process went very smoothly, especially at the beginning. However, when the time came to establish the details, viewpoints became less flexible and it took quite some effort to retain the residents' contribution. A few traffic experts thought their professional input was considerably more valuable than that of the residents, and this almost brought about the collapse of the carefully built basis of trust. There were also a few financial setbacks, which meant that certain promises had to be broken. Looking back, three phases in the process can be distinguished:

Phase 1. Uninformed optimism.
Phase 2. Informed pessimism.
Phase 3. Informed optimism.

The second phase took quite some time. However, the organisers at the municipality were quite determined and finally managed to reach phase 3 with the help of the water board. The plan is now being implemented. Perhaps the biggest surprise – especially for the municipality – was that when the process was evaluated, the residents were full of praise for it.

11. Start with Water

The process in de Vliert – and also in other Dutch residential areas – has made it clear that water has a special position in relation to other elements of the living environment. For projects aimed at the quality of the living environment, it can certainly be said that 'Water works!' Perhaps water is not people's main interest, but it does appeal as it mainly evokes positive emotions. By starting with water in this type of process, trust can be rebuilt or can grow between the involved parties. This is an investment in the social capital as it were. If there is plenty of trust, such more sensitive issues as parking, dog faeces and speed ramps can be discussed in the process. These subjects often evoke negative emotions. Water is a good 'trailblazer' with an adaptive approach as its precondition.

The example shows how technologists can interact with those from other disciplines – a process in which technology does not follow but leads the way.

12. References

1. Geldof, G.D. (1995) Adaptive water management, integrated water management on the edge of chaos, *Water Science and Technology*, **32** (1), 7-13.
2. Waldrop, M.M. (1993) *Complexity, The Emerging Science at the Edge of Order and Chaos*. Viking Books, London.
3. Tops, P.W. and Weterings, R. (1999) Een nieuwe glansrol voor de overheid? Doelmatigheid en democratie, In: *Gezichtspunten*, a CROW edition on management and maintenance of public urban space, Ede.

A UNITED KINGDOM PERSPECTIVE ON INSTITUTIONAL CONSTRAINTS LIMITING ADVANCES IN STORMWATER MANAGEMENT

R. CRABTREE
WRc plc, Swindon, SN5 8YF, UK

1. Introduction

The traditional approach to water supply and wastewater transport systems in Europe is to supply high quality potable water to the urban community and to transport the resulting 'consumed' wastewater to receiving waters beyond the margins of the urban area. The principle of centralised water supply and wastewater treatment, underpinned by high levels of investment in infrastructures and associated maintenance and renewal costs, is still largely unchallenged due to the assumed risks to health and quality of life in the city. Similarly, the traditional approach to urban expansion has been to extend the catchment area for the importation of water for potable supply and the exportation of treated wastewater. Increasing demands for clean water in the city and the resulting burden of pollution loads and ecological damage from urban wastewater can no longer be sustained without causing environmental degradation and societal disbenefits coupled to economic constraints on future growth and development. Within this context, stormwater generated by rainfall on impermeable surfaces within the city is usually regarded as an inconvenience and a potential risk to property and life; for example, as evidenced by catastrophic flooding following extreme rainfall events.

Typically, the urban manager/engineer's approach to stormwater is to develop a structural network of channels and pipes for the collection and transport of stormwater away from the city as quickly as possible. With expansion of cities over the last decades it has become common practice to separate stormwater from urban wastewater in the foul sewer system to reduce the capacity needed for foul sewers and wastewater treatment facilities. As stormwater can be rapidly generated and can contain a range of significant pollutants washed off the urban area, this approach can result in considerable degradation of downstream environments while incurring major costs in providing protection to the urban population, cultural heritage structures and other infrastructures.

While supplies of clean water were readily available from outside the city there was no incentive to treat and re-use stormwater within the city. However, as pressure on finite water resources is increasing and it is now recognised that a major proportion of the water used within a city does not need to be of potable quality, there is a potential beneficial role for stormwater. Achieving this benefit will require innovative management strategies for stormwater. A key innovation that is required is to make

J. Marsalek et al. (eds.), Advances in Urban Stormwater and Agricultural Runoff Source Controls, 305–314.
© 2001 *Kluwer Academic Publishers. Printed in the Netherlands.*

stormwater management both sustainable and optimally integrated with other urban needs.

Traditional management paradigms, including lack of understanding of system behaviour, lack of appropriate technology, and, insufficient investment in research and infrastructure have produced expensive suboptimal systems which degrade both the urban environment and downstream receiving waters. Driven by European legislation, a change in management perception has already started to occur with urban wastewater management. As a result, new wastewater technologies and holistic design and operation are becoming common place to optimise infrastructure and reduce the impact on the environment to socially and economically acceptable levels. The management of stormwater can benefit from this experience particularly as in many urban areas there are interactions between the wastewater and stormwater systems with combined sewers and combined sewer overflows. This paper reviews recent developments in the UK to improve the provision of stormwater management and identifies where current institutional constraints limit the application of sustainable, beneficial practices.

2. An Overview of Stormwater Management Practices in the UK

The last decade has seen a progressive development in technical know-how, cost-effective planning and evaluation tools for use by the UK water industry for tackling the problems associated with urban wet weather discharges through the technical framework provided by the *Urban Pollution Management Manual* [1]. This procedure has placed the UK in a position to address a backlog of improvements to wastewater treatment works and combined sewer overflows (CSOs). However, as yet, the full extent of problems associated with untreated stormwater discharges has still to be evaluated. The difficulties and frustrations of effecting changes in urban runoff management are clearly identified by reference to the current edition of the *Sewers for Adoption* manual [2]. This makes no mention at all of surface storage and infiltration systems, thereby reinforcing a policy of continued discharge of untreated stormwater to receiving waters. This has meant that developers have retained a traditional conservatism to urban drainage design. At the same time, the regulatory authorities have espoused strategies for the sustainable management of water resources and integrated catchment planning which are consistent with the UK government position on sustainable development. Key catchment planning issues contained within Regional Planning Guidance also place emphasis on sustainability principles for surface water quality [3]. Clearly, the approaches to stormwater management lag behind the scientific understanding of a source control approach to pollution management. Similarly, discharge of stormwater is the responsibility of several organisations.

In terms of urban surface water management, the diversion, attenuation and disposal of stormwater discharges at source is seen as a key concept supporting sustainability as it is focused on pre-emptive prevention of downstream water flow and quality problems, both in sewers and within receiving water bodies [4]. In addition, it can be argued that strategic, at-source disposal of surface runoff may provide a valuable contribution to aquifer storage and recovery, although no current schemes in the UK involve direct stormwater recharge. However, rising urban groundwater levels have been widely recognised as a problem and there are considerable uncertainties as to whether long-term source disposal of urban stormwater is likely to cause widespread contamination of urban groundwaters.

The impacts of stormwater discharges on receiving water systems are now well documented [5]. In contrast, there are relatively few field assessments available to evaluate actual impacts of stormwater infiltration to urban groundwaters. The potential for highway discharges to contaminate local aquifers has certainly been recognised, especially where roadside filter or fin drains outflows can directly infiltrate to underlying fissured strata [6]. The type and range of pollutants associated with urban stormwater is extremely variable, but five pollutant groups are of principal concern in terms of the potential use of urban runoff [7]. These are solids, heavy metals, hydrocarbons, pesticides and bacteria.

Within the context of integrated catchment planning, regulatory authorities are increasingly identifying a range of source control techniques which are perceived as constituting a suite of Best Management Practices (BMPs) or Sustainable Urban Drainage Systems (SUDS) for the sustainable management of intermittent urban runoff [6]. As most source control systems divert surface runoff to groundwater, it could be argued that they constitute a valuable source of aquifer recharge. The theoretical risks they present to groundwater pollution must therefore be set against the potential benefits to be gained.

3. Development and Application of SUDS

Most new development in the UK continues to be constructed with separate sewer systems as a matter of course. However, the promotion of SUDS particularly by the Scottish Environment Protection Agency [8] and the Environment Agency in England and Wales [9] is a key factor in promoting the uptake of alternative approaches by developers and builders. These include source control and soft engineered treatment techniques to modify traditionally engineered pipe-based approaches for the attenuation and disposal of runoff. All developments require consideration of surface water drainage. This should be incorporated into development plans which require planning consent. The UK's Town and Country Planning System is operated by local authorities supported by national planning policy guidance statements and regional/local development plans to prevent future problems arising from new development [10]. The relevant environmental regulator and sewerage undertaker will normally advise the planning authority on appropriate surface water drainage systems and techniques. Currently, larger surface water discharges may require a consent to discharge by the environmental regulator. Normally this would be some form of qualitative emission standard based on a design related to perceived current good engineering practice. Surface water discharges to infiltration systems, including soakaways, are subject to control under water pollution and groundwater regulations if polluting substances are present in the discharge; for example, from highways, parking areas and industrial sites.

SUDS are promoted as a concept that focuses decisions about drainage design, construction and maintenance on the quality of the receiving water and surrounding urban environment. As physical structures, SUDS include ponds, wetlands, swales and porous surfaces to provide attenuation and treatment of runoff. The variety of options available allows the design to be fitted to local circumstances as well as to account for more traditional design criteria, such as peak flow and storage capacity. The advantages of SUDS are recognised by most organisations involved in the planning process and offer a Best Practicable Environmental Option [11] that is entirely compatible with the requirements for sustainability. The key benefits are:

- protect and enhance biodiversity in receiving waters;
- maintain or restore the natural flow regime in urban streams;
- provide protection from flooding;
- protect receiving waters from pollution;
- extend the capacity of existing sewerage systems;
- provide amenity benefits;
- allow for groundwater recharge;
- simplify construction; and,
- reduce construction costs.

Clearly, SUDS techniques, while relatively new to the UK, are widely practised elsewhere in the world. SUDS are widely promoted by environmental regulators and planning authorities, particularly for flow control and flooding protection after development of greenfield sites. However, construction and building regulations continue to be based on the use of traditional 'hard' drainage approaches. Construction methods lag behind many of new 'soft' engineering developments that underpin some of the benefits of SUDS approaches, particularly in relation to infiltration techniques, the use of soil and vegetation in construction and green landscaping. This position is beginning to improve; for example, through the revision of the Building Regulations to provide guidance on the design and construction of sustainable urban drainage systems [12]. However, this process will inevitably take a long time and acts as a constraint on developers and builders. A further significant constraint is on the legal position of SUDS based drainage in relation to subsequent ownership and maintenance responsibilities post development. This is a complex legal area, including public safety issues, and one in which sewerage undertakers, whose main area of activity is the provision of wastewater transport and treatment, may have little incentive to adopt non-traditional stormwater drainage systems. This is particularly the case with the privatised sewerage undertakers in England and Wales. In Scotland and Northern Ireland, where the sewerage undertakers remain as public bodies, and legislation is more clear, there have been major projects based on the use of SUDS and a guidance manual has been produced to assist developers and planners [13]. A similar manual for England and Wales has also been produced [14].

4. Do SUDS provide all the answers?

There are clearly appreciated benefits in using source control and treatment techniques for stormwater management to improve the sustainability of stormwater infrastructures. However, two areas of concern are beginning to appear in the UK, and elsewhere, based on experience of the SUDS approach. The first is the ability to provide satisfactory protection to the quality of the receiving water ecosystem in terms of the quality of flow. The second is the degree of sustainability of SUDS, which remains to be identified and quantified. While the SUDS approach is better than traditional approaches in many cases, it may not provide an optimum solution as promoted by some advocates.

4.1 RECEIVING WATER PROTECTION

As originally conceived, the prime function of stormwater management is to provide protection from flooding and reduce peak flows. Sedimentation, adsorption and biological degradation of pollutants can also be achieved by appropriate selection of designs. This pollution control capability is also being actively promoted by the environmental regulators. Table 1 presents guidance provided by the Environment Agency on the pollutant removal capacity of a selection of SUDS designs [9].

TABLE 1. Pollutant removal capacity for selected SUDS design [9]

SUDS Design	Pollutant removal (%)			
	Solids	Nutrients	Metals	BOD
Grass swale	30	30	10	30
Porous pavement	100	80	100	100
Infiltration basin	100	80	100	100
Wet pond	100	50	50	80
Detention basin with vegetative treatment	100	80	80	60
Detention basin	80	30	50	30

While those designs can achieve significant levels of pollution removal, the resulting concentrations in the discharge will be related to the initial concentrations, which are known to be highly variable in relation to land use and event characteristics. Despite major research efforts it has not been possible to provide reliable methods for predicting stormwater pollutant concentrations [7, 15]. Hence, while the impact of a discharge from a SUDS based drainage system should be less than that from a traditional system in similar circumstances, it may not be acceptable in terms of receiving water quality and ecosystem impact. In some cases, it appears that the ecological impact of the discharges from traditional systems and SUDS may be comparable. To address these issues, the discharge has to be related to the assimilative capacity of the receiving water.

An approach based on the achievement of receiving water quality standards to enable beneficial uses to be met has been the accepted UK practise, as required by the environmental regulators' policy, for controlling urban wastewater discharges for a number of years. Major combined sewer overflow improvement programmes are in place based on this approach, termed Urban Pollution Management, and commonly referred to as UPM [1, 16]. This is underpinned by the use of wet weather specific environmental criteria applied to receiving waters to control the impact of discharges to an acceptable level. These criteria [16] are designed to protect river ecosystems from the effects of intermittent wet weather urban discharges. Table 2 illustrates the Dissolved Oxygen criteria currently employed by the Environment Agency in England and Wales [16]. Similar criteria, based on a concentration/duration/frequency concept, are also applied for un-ionised ammonia. Both sets of criteria are designed to avoid fish deaths for pollution episodes up to a return period of 1 year. The environmental quality standards approach allows integration of the design and management of the wastewater system to minimise costs and achieve the necessary environment benefits. Clearly, the current emission criteria approach applied to stormwater discharges is at odds with the management of urban wastewater.

TABLE 2 Fundamental Intermittent Standards for Dissolved Oxygen - concentration/duration thresholds not to be breached more frequently than shown [16]

a) Ecosystem suitable for sustainable salmonid fishery

Return period	Dissolved Oxygen concentrations (mg/l)		
	1 hour	6 hours	24 hours
1 month	5.0	5.5	6.0
3 months	4.5	5.0	5.5
1 year	4.0	4.5	5.0

b) Ecosystem suitable for sustainable cyprinid fishery

Return period	Dissolved Oxygen concentrations (mg/l)		
	1 hour	6 hours	24 hours
1 month	4.0	5.0	5.5
3 months	3.5	4.5	5.0
1 year	3.0	4.0	4.5

c) Marginal cyprinid fishery ecosystem

Return period	Dissolved Oxygen concentrations (mg/l)		
	1 hour	6 hours	24 hours
1 month	3.0	3.5	4.0
3 months	2.5	3.0	3.5
1 year	2.0	2.5	3.0

Notes
1. These limits apply when the concurrent un-ionised ammonia (NH3-N) concentration is below 0.02 mg/l. The following correction factors apply at higher concurrent un-ionised ammonia concentrations:
   0.02 - 0.15 mg NH3-N/l: correction factor = + (0.97 x loge(mg NH3-N/l) + 3.8) mg O/l
   >0.15 mg NH3-N/l: correction factor = +2 mg O/l.
2. A correction factor of 3 mg O/l is added for salmonid spawning grounds.

At present much environmental legislation remains single purpose. However, driven by EU Directives, changes are being made to make environmental legislation more integrated and enable more holistic, sustainable approaches to be implemented. In particular, the forthcoming Water Framework Directive is based on the principles of sustainable development. How this will be translated into UK legislation is, at present, unclear but will be based on:

- river catchment based planning;
- achievement of good ecological status for surface waters;
- protection of groundwaters;
- pollution control by environmental quality standards and emission limits; and,
- consideration of physical, quantity and quality characteristics for surface waters.

The impact of these criteria on stormwater management will result in the need to integrate stormwater management and urban wastewater management to enable receiving water quantity, quality and ecological criteria to be met cost-effectively. As a result, the future approach for consenting significant stormwater discharges is likely to be consistent with the policies applied to wastewater system wet weather discharges. Ultimately, the benefit will be improved environmental performance by integrating the control of all urban wet weather discharges and more sustainable urban infrastructures.

4.2 SUSTAINABILITY

SUDS based systems are likely to be more sustainable than traditional systems in many cases. However, sustainability covers many issues, from design and construction through to full life cycle assessment. At present these issues have only been considered at a subjective, qualitative level. To date, no assessment has been made to identify sustainability indicators and performance levels for urban drainage systems. These must be derived before the true, underlying sustainability of stormwater management designs and practices can be established and placed within the context of the overall sustainability of an increasingly urbanised society. While these wider issues are beyond the scope of this review, potential sustainability criteria for stormwater management are presented in Table 3.

5. The Way Forward

While the UK is often looked to by others as leading the way in river basin management the uptake of new ideas is not a strong trait. There have been two major research initiatives in the last decade aimed at improving the management of urban wastewater and stormwater systems. The first of these is UPM and the second is SUDS. However, the implementation of SUDS approaches for urban stormwater is a reflection of a traditionalist approach to urban water management. There is a general lack of knowledge and appreciation of the performance of SUDS concepts within the wider context of urban development and environmental protection at a regional and catchment scale. In particular, there is a focus on the disposal aspect of stormwater management rather than its reuse as a beneficial resource. Current institutional arrangements for stormwater management in the UK are seen as a major potential constraint to the introduction and uptake of sustainable and beneficial stormwater management through the use of appropriate combinations of technologies that are integrated with the wider urban infrastructure.

These two important approaches, SUDS and UPM, have been developed in relative isolation of one another, yet they are potentially compatible techniques for managing the impacts of urbanisation. However, little thought has been given to the potential synergistic benefits of fully integrating the two approaches. SUDS can have a role to play in solving the typical problems addressed in UPM studies aimed at rehabilitation of large scale systems. Also SUDS techniques can be represented in the UPM modelling framework. An approach which allows the benefits of both techniques to be fully achieved is likely to be adopted in the future.

TABLE 3. Potential sustainability criteria for stormwater management

	Area	Relevant issues
1.	Ecological Impact	- runoff peaks - river flow and groundwater recharge - sedimentation - release of pollutants - interaction with urban wastewater - soils, fauna and flora
2.	Construction Operation and Maintenance	- efficiency - use of materials and energy - renewable consumption - flood control - local sources of labour and materials - integration with urban infrastructure - adaptable to change
3.	Social and Urban	- health and safety - protect property - cultural heritage - visual impact - multiple use

Different approaches to stormwater management exist and are being developed within Europe and beyond, particularly in North America. New strategies and novel technologies are being introduced which seek to achieve sustainability requirements and achieve the traditional needs of flood protection and natural hazard limitation. The potential effects of climate change on urban areas also provides a driving force and opportunity to increase the benefits of stormwater for retention and re-use in the urban water cycle. Future integrated sustainable stormwater management strategies must consider the following tenets for system design and operation:

- protection from flooding by reducing peak rates of urban runoff;
- reduction of pollution loads to minimise environmental impact within and downstream of the urban area;
- protection of groundwater recharge to sustain baseflows in urban streams;

- collection and retention of stormwater for re-use as a beneficial resource within the urban area;
- use of stormwater transport and treatment system components as water features to enhance the urban landscape; and,
- reduction in the use of natural, economic and cultural resources in the construction, operation and maintenance of stormwater infrastructures.

To some extent these tenets are being considered in stormwater system design through the introduction of SUDS concepts such as source control for flow and pollution reduction coupled with end of pipe control devices; the opening out of culverted channels within the urban area; recharge of aquifers from stormwater collection ponds; soil bioengineering construction and landscape planning, and, educational programmes. However, in the UK, there is a general lack of knowledge and appreciation of the actual performance and benefits of such approaches as part of the wider context of urban planning and development. Hence, there is a need to exchange experience and knowledge to identify and promote best practices for sustainable and beneficial stormwater and wastewater management.

6. References

1. FWR. (1994) *Urban Pollution Management Manual*, Foundation for Water Research, Marlow.
2. WRc. (1995) *Sewers for Adoption, 4th Edition*, WRc, Swindon.
3. NRA. (1993) Thames 21: *A planning perspective and a sustainable strategy for the Thames Region*. National Rivers Authority, Thames Region, Reading.
4. SEPA. (1996) *A Guide to Surface Water Management Practices*. Scottish Environmental Protection Agency, Stirling.
5. Ellis, J.B. and Hvitved-Jacobsen, T. (1996) Urban drainage impacts on receiving waters. *Journal of Hydraulics Research* 34(6), 771-783.
6. CIRIA. (1994) *Control of Pollution from Highway Drainage Discharges*. Report No. 142 Construction Industry Research and Information Association, London.
7. Ellis, J.B. and Crabtree, R.W. (1999) Organisational Issues and Policy Directions for Urban pollution Management, in. S.T. Trudgill, D.E. Walling and B.W. Webb (eds) *Water Quality: Processes and Policy*. John Wiley & Sons Ltd. pp 181-200.
8. SEPA. (1999) *Sustainable Urban Drainage: An Introduction*. Scottish Environment Protection Agency, Sterling.
9. Environment Agency. (1999) *Sustainable Drainage Systems: A guide for developers. Interim Advice Note C10*. Environment Agency for England and Wales. Bristol.
10. DETR (2000) *New Planning Policy Guidance Note 25:Development and Flood Risk: Consultation Draft*. Department for Environment Transport and the Regions.
11. WRc. (1998) *Control of Pollution from separately sewered areas: Manual of Good Practice*. Report No. PT2039. WRc, Swindon.
12. DETR. (2000) *The Building Act 1984 – Review of Part H (Drainage and Solid Waste) of the Building Regulations 1991 and associated legislation*. Department for Environment, Transport and the Regions.
13. CIRIA. (2000) *Sustainable urban drainage systems – design manual for Scotland and Northern Ireland*. Construction Industry Research and Information Association, London.
14. CIRIA. (2000) *SUDS Design Manual for England and Wales*. Construction Industry Research and Information Association, London.
15. Crabtree, R.W., Garsdal, H., Gent, R.J. and Dorge, J. (1994) MOUSETRAP: a deterministic sewer flow quality model, *Water Science & Technology* 30(1), 107-115.
16. FWR. (1998) *Urban Pollution Management Manual – 2nd Edition*. Foundation for Water Research, Marlow.

7. Acknowledgements

This paper has been produced with the permission of the Directors of WRc. The views expressed in the paper are those of the author and not necessarily those of WRc.

INDEX